THE AUTHOR

Michael Cooley was born in Tuam in the west of Ireland in 1934. He was educated at local Catholic schools and later studied engineering in the UK. In industry he specialised in engineering design and gained a PhD in computer-aided design.

Mike Cooley was national president of the Designers' Union in 1971 and a TUC delegate for many years. A design engineer for eighteen years, he was a founder member of the Lucas Aerospace Combine Shop Stewards' Committee and one of the authors of its Plan for Socially Useful Production.

He has lectured at universities in Australia, Europe and the United States. He has been a guest professor at the University of Bremen, and visiting professor at the University of Manchester Institute of Science and Technology. He has written for a variety of publications world wide including the *Guardian*, the *Listener* and the *New Statesman*. He has produced over forty scientific papers and is author or joint author of eleven books in English and German and has contributed to some thirty-five more. His work has been translated into over twenty languages from Finnish to Japanese. He is an international authority on human-centred computer-based systems and in 1981 was joint winner of the $50,000 Alternative Nobel Prize, which he donated to the Lucas Combine Committee.

Mike Cooley has been chairman and director of several manufacturing companies in his capacity as director of technology of the Greater London Enterprise Board in the 1980s.

His book, *Arhcitect or Bee? The Human Price of Technology* was re-published in 2016 and *Delinquent Genius: The Strange Affair of Man and His Technology*, written in 2008, was published in 2018 by Spokesman.

"We are the only species ever to have it within its power to destroy itself along with our beautiful and frail planet. This is an awesome capability and one for which our culture, education and politics ill prepares us to cope creatively."

Mike Cooley
From judgement to calculation

The Search for Alternatives

Liberating Human Imagination
A Mike Cooley Reader

SPOKESMAN
Nottingham

First published in 2020

Spokesman
5 Churchill Park
Nottingham, NG4 2HF, England
www.spokesmanbooks.com

Spokesman is the publishing imprint of the Bertrand Russell Peace Foundation

© Michael Cooley

All rights reserved. Except for brief quotations in a review, this book, or any part thereof, may not be reproduced, stored in or introduced into a retrieval system, or transmitted, in any form or by any means, electronic, mechanical, photocopying, recording or otherwise, without the prior written permission of the publisher.

ISBN 978 085124 885 1

A cataloguing-in-publication (CIP) record is available from the British Library

Printed in the European Union

Contents

Foreword by John Palmer 1
Introduction by Karamjit S Gill 5

1. Computer Aided Design:
   Its Nature and Implications 9

2. Contradication of Science and Technology
   in the Productive Process 47

3. The Search for Alternatives 77

4. Science and Social Action 87

5. The Lucas Plan: An Interview 103

6. Human Centred Systems 121

7. Technology and International
   Development 137

8. The Myth of the Moral Neutrality
   of Technology 161

9. My Education 173

10. From Judgement to Calculation 181

Foreword

John Palmer

Prophets and prophecy have not always had a good press. Too often prophetic visions have had limited relevance for those in society and especially the world of work confronting real and immediate challenges. Prophets with their feet placed firmly on the ground and, specifically, those concerned with the so-called 'mundane' world of work have been very rare indeed.

Mike Cooley has every reason to be counted as one such 'prophet.' He has rightly been described by the President of Ireland, Michael D. Higgins, as "… the most intelligent Irish man, the most morally engaged scientist and technologist Ireland has sent abroad." He is certainly a visionary of a future where human skill and labour work in partnership with science and technology rather than in servitude to them.

Born in Tuam, County Galway in Ireland, Mike Cooley studied advanced computer based engineering in Germany and Switzerland. He first came to public attention as a result of his pioneering role in the British trade union movement as an advocate of human skill being enhanced by and not harnessed or displaced by technology.

As a trade unionist working in the Lucas Aerospace company he played a key role in outlining how workers could confront the threat of mass redundancies by showing how their skills could be adapted to produce alternative "socially useful" products and demonstrated practical examples in health, transport and other sectors.

The socially responsive ethos of the human-centred movement generated by the Lucas Workers' Plan of 1976 is summed up in the Mike's statement that "there cannot be islands of social

The Search for Alternatives

responsibility in a sea of depravity". He also warned about "the appalling gap between what technology could provide for society, and what it actually does provide."

> "The tragic waste our society makes of its most precious asset— the skills, ingenuity, energy, creativity and enthusiasm of ordinary people'; and 'the myth that computerisation, automation and use of robotic devices will automatically free human being from soul destroying, backbreaking tasks and leave them free to engage in more creative work".

Mike Cooley's vision was of a human-machine symbiosis as an alternative potential for work life. He saw this as part of wider European humanistic movements such as 'Democratic Participation' (Scandinavia) and 'Humanisation of Technology and Work' (Germany). These European human-centred movements provided a basis for the establishment of the 'Anthropocentric Systems and Technology' programme of the European Union during the 1990s.

If his approach had received the political support in the wider labour movement it deserved, perhaps the worst depredations of Thatcherite 'slash and burn" economics in the 1970s and 80s might have been mitigated or avoided. In later years his role as Director of Technology in the innovative Greater London Enterprise Board (GLEB) allowed these ideas to reach a wider audience.

But its work was undermined when Margaret Thatcher's government closed the Greater London Council which had sponsored the creation of the GLEB. Alas the longer term benefits of the radical strategies Cooley and others canvassed were more often realised in other European countries rather than the UK.

It was left to others to invest in projects such as a pioneering 'road/rail bus' and a new type of portable kidney machine. We are still paying the price today in the post-industrial desert created by the 'free' market, particularly in the former industrial areas of Britain. The economic deprivation and social inequality generated by the free market system has spawned new forms of extreme right populism.

Foreword

Mike Cooley has also played a crucial role in developing thinking about how the interplay between the diversity of human skill and the calculation capacity of the machine can lead to enhanced productivity and enriched human expertise, combining human ingenuity and technological innovations.

Cooley has warned us of the danger of the objectification of human knowledge and experience into information and data, risking human judgement becoming mere calculation and turning the human into a mere robot. This was dramatically expressed in his pioneering book *Architect or Bee.*

Alas neo-liberal capitalism has taken us down a very different path. The monumental squandering of the creative potential of working people – in partnership with human centred science and technology – has led to the emergence of casualised labour in both industries and services. Far from productivity being given a massive boost, too many people work in low productivity sectors, low paid and often without adequate protection from unemployment and fluctuating income.

When Mike talked about the complex and little understood relationship between a worker's innate skills and the tools and technologies available to them, he would draw on a rich knowledge of history. "Think of the breath taking achievements of the mediaeval workers who built the great cathedrals. Who were the architects?" he heard him ask.

His response was clear:

"Actually there was no separate cast of architects giving instructions to a passive work force. Rather every stonemason and building worker had an innate sense of the potential design born from long and intimate knowledge of the materials with which they were working".

The impressive scope of Cooley's thinking is well reflected in his subsequent books *The Human Price of Technology* republished in 2016 and *Delinquent Genius: The Strange Affair of Man and His Technology* which was published in 2018 by Spokesman. He was

The Search for Alternatives

quick to identify the gender bias in the pattern of contemporary work and skills distribution.

The thrust of Mike Cooley's human society focused analysis has a striking parallel in the rapidly growing world-wide movement – led by young people – against climate change and for radical, green policies in all the major aspects of our economic, social and individual lives to counteract it.

It is a development profoundly welcomed by him as the climate change threat to our planet and its people looms ever larger. In a sustainable world economy, the values and goals of Michael Cooley's work on human centred technology are sure to be reflected.

It is encouraging to know that Mike's massive contribution to new thinking about a human centred economy and society will be available for future generations to draw inspiration from. The Waterford Institute of Technology's (WIT) INSYTE Research Centre in Ireland has acquired by donation from the Cooley family the entire archive of Prof Mike Cooley. The Mike Cooley archive consists of 1,400 items which will be digitised and in the long term and made available online.

John Palmer, December 2019
John Palmer was Public Affairs Director of the
Greater London Enterprise Board and is a
former European Editor of *The Guardian*

INTRODUCTION

Karamjit S Gill

Through his historical insights into the evolution of digital technology, from calculation to computation of the recent past, Mike Cooley provides a stimulating narrative for understanding the impacts and implications of the new digital technologies of machine intelligence and automation. His argument is that only by gaining insight into the historical evolution and context, can we "identify discernible laws of development, and having understood these laws to use them to scientifically predict what effects the equipment will have ... in the future." Cooley's main concern is the misuse of technology, which can amongst other things create a frantic work tempo for some and the dole queue for others. He recognises that technology and social organisation interact to elevate the nature of human existence to a higher level, whilst appreciating that even the most sensitive human faculty, that of memory and the nervous system, has now in many ways been extended by computer supported decision-making. Although humans with their skill and ingenuity were able to create technological change from the early stages to the advance of artificial intelligence, the society which has given birth to them tends to fail to keep pace.

This volume is an important asset as we seek critical approaches to make sense of the data driven society of the 21st Century, whilst grappling with the impact of automation on the one hand, and envisioning the common-good potential of augmented artificial intelligence (AI) systems on the other. We face social challenges of governance, ethics, accountability and intervention arising from the accelerated integration of powerful artificial intelligence

The Search for Alternatives

systems into core social institutions. The automation agenda of the work place continues to happen quietly out of public view, hidden behind the public mantras of "digital society", "human-centered A.I." and the "Fourth Industrial Revolution". With the exponential rise of big data flows in networked communications and their manipulating algorithms, the gaps in translation are now too vast to grasp and address, rendering us unable to engage with difference through the shadows of machine thinking. Augmentation and automation places the human in the predicament of accepting the calculation of the machine without judgment. We echo Cooley's concerns of 'socially irresponsible' science and ask whether we can transcend the instrumental reason of machine thinking to mould technological futures for the common good rather than turning them into a single story of 'singularity'. Can we re-appropriate the idea of causality that has been taken by 'science' and reframe it in the making of everyday judgments and decisions? How can we harness collective intelligence as a transformational tool for addressing complex social problems? Just as Cooley narrates his argument situated in the context of his days, we need to draw upon various AI narratives of the relations between society and the scientific project of AI and the challenges it poses for us to come up with possible symbiotic AI futures.

In exploring AI futures, we should take note of Cooley's reminder that the scientific project is always embedded within a particular social order and reflects the norms and ideology of that social order. In this perspective, science ceases to be seen as autonomous, as it internalises ideological assumptions, thereby shaping the design of the systems, tools and theoretical frameworks of its validation. Cooley notes that throughout history, science has shaped the ideas and critical knowledge that contributed to liberating humanity from the bondage of superstition and religion that acted as a key ideological prop of the outgoing social order. Darwin, in making redundant earlier ideas of the creation of life and of humanity, and the Galilean revolution which destroyed the earthcentred model of the universe, illustrate science is not just shaped by the ideological assumptions but also

Introduction

shapes the rationalities that are practised by society. Cooley thus makes us aware that critical oversight of the emerging technologies of the artificial extends far beyond that of scientific abuses, to deeper considerations of the nature of the scientific process itself.

Cooley notes that it is true that the drive for scientific knowledge has provided the material basis for a fuller and more dignified existence for the community as a whole. It must not, however, be a blind and unthinking drive forward, shirking our social and political responsibility to analyse its effects upon society. Any meaningful analysis of scientific abuse must probe the very nature of the scientific process itself, and the objective role of science within the ideological framework of a given society. As such, it ceases to be merely a 'problem of science' and takes on a political dimension. It extends beyond the idea of important, but limited, introverted soulsearching of the scientific community, and recognises the need for wider public involvement.

Just as the old technology arrived at a historical breaking point at which the old societies were deemed to be transformed into a new one, the technologies of the artificial are now beginning to generate a situation in which society once again is facing the spectre of a new transformation. The challenge is to create a strategic framework that facilitates this change, in response to the technologies of computerization and automation, for example in dealing with the disruption of social, economic and cultural life, especially when life becomes synchronised with the computerised environment. Cooley asserts that the rise of contradictions in technology and society "cannot be resolved within the framework of a free enterprise system, since they are but manifestations of the irreconcilable contradictions between the interests of the exploiter and the exploited." For Cooley, 'socially irresponsible' science not only pollutes our environment, it also degrades us, both mentally and physically, as mere objectified beings and reduces us to mere machine appendages.

Karamjit S Gill
Editor, *AI&Society*
Cambridge, UK

COMPUTER AIDED DESIGN
ITS NATURE AND IMPLICATIONS

First published by the AUEW (TASS), 1972

PREFACE

This booklet is based on lectures delivered by the author during 1970/71. Those sections dealing with the social, industrial and political implications of technological change and the history of computerisation are extracted from lectures delivered at the TASS Summer School in Ruskin College, Oxford, June 1971, and a number of Weekend Schools.

INTRODUCTION

The mystique and jargon which surrounds the use of computers almost surpasses the feudal mysticism which is still evident in the medical profession. Even technicians are likely to be intimidated by some of the formidable expressions used by computer personnel. Many TASS members with a life-time of practical design experience, backed in many cases by academic qualifications, still feel themselves threatened by these new techniques.

The techniques in themselves represent no threat. The threat arises from the manner in which these techniques are applied. Recognising that this is so, and that computer aided design is an inevitable and growing part of everyday industrial life, the Executive Committee decided that I should prepare a booklet on the subject. The object of the booklet is twofold. Firstly, to cut across the jargon and explain in everyday terms the nature of

The Search for Alternatives

computer aided design, with examples of actual applications, and secondly, to draw to the attention of our members and the rest of the trade union movement the employment and occupational dangers which the uncontrolled introduction of these techniques in a profit-orientated society can bring in their wake.

The booklet takes the form of three self-contained sections, 'Historical Background', 'Computer Aided Design Examples' [not published here] and 'Technological Change, Its Effects'.

The first section traces the evolution of the methods of calculation from earliest time to the introduction of computers in their present form. It should be of general interest even to those not directly involved in the design field.

The second section provides some typical examples of computer aided design applications. These have been drawn from a wide spectrum of industry and, since they are of necessity somewhat technical, interest in this section is likely to be limited to those involved in design and allied technical fields.

The third section looks at the human, social and industrial consequences of computer aided design. It recognises that design staff are not an island unto themselves, and that their destiny, conditions and problems are bound up with the rest of the working class. Hence the consequences of computer aided design are analysed in the broader context of the consequences of technological change as a whole. This is probably the most important section in the book, since in the author's view people are more important than machines. (Although that may sound obvious it is a proposition that is day by day challenged in practice by the monopolies which dominate our economy). This section should be of direct interest to all active trade unionists, irrespective of their occupation or industry. It is hoped that it will provide the basis of a far-reaching discussion for those at the point of production – the people who have the real power to determine whether technological change will be used in their interest or to their detriment.

CAD – Its Nature and Implications

Section 1
HISTORICAL BACKGROUND

It is frequently felt that computerial techniques have only been used during the past ten years. In fact, the techniques used today are a logical progression of the use man has made of the science of numbers for thousands of years. Even before the dawn of recorded history numbers and measurement formed an important part of commercial dealings. These early computations dealt only with simple mathematical techniques such as addition, subtraction, division and multiplication.

The great stimulus for the science of numbers came, however, not from man's commercial transactions but from his desire to understand the universe in which he lived. Man's interest in the stars and their movements, and in the related subject of navigation, provided the need for more complicated calculations. Once having evolved such methods of calculation man's ingenuity induced him to devise methods by which the human brain could be relieved of the routine tasks involved.

We can reasonably suppose that early man used the fingers of his two hands to represent numbers. This, however, did limit the counting up to ten. The next development, said to be still in use in Africa, is to enlist the aid of a second man. The first man counts the units up to ten on his fingers, while his partner counts the groups of ten so formed. The development from this, which took place in the first civilisations of Egypt, was to represent the numbers by means of paddles arranged in heaps of ten.

The Abacus

This technique gradually led to the development of the abacus or counting frame (see Fig. 1). It was widely used throughout the Mediterranean world in the first millenium BC. The abacus is still very widely used in China and Japan. The Japanese abacus in the hands of a skilled operator is an extremely impressive instrument. Indeed, as late as 1946, in a contest between a Japanese specialist and an American using the then most modern electrical desk calculating machine, the abacus won a contest covering five types of calculations judged on speed and accuracy.

THE ASTROLABE AND SLIDE RULE

The next important calculating machine to be invented was the astrolabe of Greek origin, invented about 2000 BC. It was used by Arab seamen, and its introduction gave a sound basis for nautical surveying. In fact, this instrument was in use up to the middle of the 18th century, when it was superseded by John Hadley's reflecting quadrant or sextant. The astrolabe was used to measure the positions of the sun, the moon and the stars in the sky. From these measurements it was possible to compute the latitude at any place. The astrolabe had two sides, one for measurement and one for calculation. Thus it was capable of collecting the information and processing it. Fig. 2 shows the calculation side.

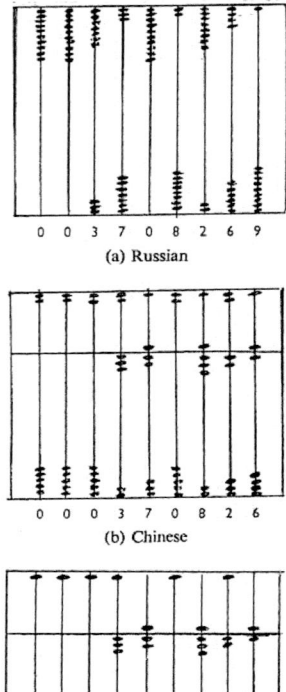

Fig. 1

John Napier invented logarithms in 1614. This in turn led William Oughtred in 1621 to devise a pair of sliding scales upon which multiplication and division could be carried out. That instrument is still very much in use today and is, of course, the slide rule.

CAD – Its Nature and Implications

Fig. 2
An astrolabe

MECHANICAL CALCULATING

The invention of the first true mechanical calculating machine is usually attributed to the philosopher and mathematician, Blaise Pascal. Working to his own drawings Pascal completed his first machine in 1642. In presenting his machine to the public he said:

> "I submit to the public a small machine of my own invention, by means of which you alone may, without any effort, perform all the operations of arithmetic and may be relieved of the work which has so often fatigued your spirit when you have worked with the counters or with the pen."

Some of Pascal's models are still preserved in Paris.

It was not, however, until about 1810 that the first commerical calculating machines were sold. About 1,500 of these were marketed. The basic principle, as in Pascal's machine, was a gear mechanism. It was evident at this stage that any further advance had to involve an automatic method of feeding in the numbers.

The Search for Alternatives

PUNCHED CARDS

The method by which this could be done was evolved by Joseph Marie Jacquard, 1752 to 1833. He designed a loom in which the threads were controlled by means of punched cards. Each card was punched with a pattern of holes and arranged to pass over a drum in sequence on a string. During each throw of the shuttle one of the cards, the next in sequence, pressed against the whole set of loom rods. Those rods which were opposite a hole in the card remained at rest; the other rods were lifted.

Thus, at the beginning of the 19th century it was possible to design a pattern and print the information on punched cards, and then have that pattern woven on a Jacquard loom. By using sufficient cards it was possible to design a most complex pattern. A portrait of Jacquard himself used 24,000 punched cards, and so fine were the details that most people took the finished product to be an engraving. Thus it will be seen that some 150 years ago the basic technique of punched card preparation of information was evolved and this is still the basic technique in use today.

DIFFERENCE ENGINE

The next major advance was made by an Englishman, Charles Babbage. He produced and demonstrated his Difference Engine in 1822 – See Fig. 3. The machine is still to be seen in the London Science Museum. Ten years later, however, Babbage embarked upon a more ambitious scheme, which he called the analytical engine. This was to all intents and purposes a universal computer, and embodied in it many of the techniques used even to this day. Although the design concepts behind the Analytical Engine were completely sound, the basic trouble was that the scheme was far too ambitious. For example, it was his idea to work with numbers to 50 decimal places. The level of precision engineering in the early 19th century was totally inadequate to meet his demands. In consequence of this the machine never became truly operational. It did, however, embody many of the techniques still in use.

Fig. 3
A portion of Babbage's Difference Engine
Crown copyright, Science Museum, London.

ELECTRO-MECHANICAL CALCULATING

A further advance was made some 20 years after Babbage's death. This was the invention of the electro-mechanical punch card calculating machine by Herman Hollerith. Hollerith was a statistician on the staff of the US Bureau of Census. In 1886 the returns of the 1880 US census were still being counted and sorted, and it was clear that with the methods then existing the job would

The Search for Alternatives

be unfinished in 1890, when the next census was due. Hollerith saw that the solution lay in some measure of mechanisation, and set about the task of devising suitable equipment. He based this upon the punched card system used in the Jacquard looms. Some of the machines he devised were used in the US Census of 1890. The first punched cards were used in Britain on the Census of 1911.

The next major development was in 1937 when Howard N. Aiken of Harvard had the idea of using the techniques and components developed for punch machines to produce a fully automatic calculating machine. He approached International Business Machines Corporation (IBM), and the result of their joint collaboration was the automatic sequence controlled calculator, which was completed in 1944. The completed machine was extremely large, cumbersome and slow by present day standards, yet laid the basis for modern techniques. The completed machine contained more than three-quarters of a million parts, and used more than 500 miles of wire. Addition or subtraction of two numbers took three-tenths of a second, multiplication about four seconds, and division about ten seconds. Its real claim to fame lies in the fact that it was the first fully automatic computer ever to be completed, and laid the basis for the highly complex machines that are now in use.

ELECTRONIC TECHNIQUES

Thereafter, the development of computers was extremely rapid. Electronic techniques were applied to computer design in 1946 when the famous ENIAC (Electronic Numerical Integrator and Calculator) was completed. This machine was designed primarily to meet specialised military needs; for example the calculation of the trajectory of bombs and shells. By describing this as the first electronic computer, it is meant that the storage and multiplication of numbers inside the machine, and also the control of the sequence of operation, was done by means of electronic circuits. Indeed, apart from the input and output mechanisms, this machine

CAD – Its Nature and Implications

had no moving parts. The electronic techniques also enabled the operating speed to be increased enormously. It was capable of adding numbers at a rate of 5,000 per second, but it did contain something like 18,000 valves.

The development from this stage was so rapid and so diverse as to require a book in itself. It is, however, worth mentioning that the first electronic computer in Britain was developed at Cambridge in 1949, and the first commercial computer was designed and developed by Ferrantis in 1951.

This historical background, although sketchy, will serve to illustrate that computers and the use of machines to free man from routine calculations is not a technique which has suddenly been sprung upon us, but is part of a long technological evolution. That which differs today is the rate at which these techniques are being applied.

LANGUAGES

Computer languages may for convenience be divided into two major groups. Firstly, there are commercial languages which are used for accounting and business routines. Secondly, there are the scientific problem orientated languages. These are used where a number of variables are involved, and are applied mainly to scientific and mathematical work.

A typical commercial language is Cobol – Common Business Orientated Language. The instructions in this language are given in stylised English. It includes instruction words such as "add" and "to". It also includes a number of stylised English labels such as "overtime", "normal hours", etc. It can give the computer direct instructions in this stylised form of English, such as "If stock less than minimum go to recorder". The advantages of this language are that it is extremely simple to learn and it is possible to read other people's programmes. Its major disadvantage is that the instructions which can be given are very limited, and are appropriate only to business routines.

TASS members in their work in the design environment are far

The Search for Alternatives

more likely to encounter the scientific problem orientated languages. Two typical examples are 'Fortran' and 'Algol'. Fortran obtains its name from Formulae Translation. It is built up of five basic operators. +(add), -(sub), *(multi.), /(div) **(exponentiation). An example of the well known formula for one root of a quadratic equation written in Fortran would be as follows:

$$ROOT = (-B + SQRT(B**2-4*A*C))/2*A$$

It will be seen that the language is comprised of a combination of algebraic formulae and English language statements. For example in the formula above the machine is capable of translating square root etc. into the appropriate machine code.

Algol derives its name from Algorithmic Language. Like Fortran it is a problem orientated, high level programming language for mathematical and scientific use, for which the source programme provides a means of defining algorithms as a series of statements and declarations, having a general resemblance to algebraic formulae and English sentences. An Algol programme consists of data items, statements and declarations organised in a programme structure in which the statements are combined to form compound statements and blocks. As in Fortran, numbers and a series of standard functions are combined by five operators. By way of comparison, one root of a quadratic equation is given below in Algol:

$$ROOT := (-b + SQRT(bt\ 2-4 \times ax\ c))/(2 \times a);$$

BINARY NOTATION

Many older traditioaal draughtsmen, when confronted with technological change in the design environment, find the terms and methods somewhat confusing. One of the systems which sometimes creates problems is the requirement to use binary numbers. If binary numbers are seen as a logical extension of the

CAD – Its Nature and Implications

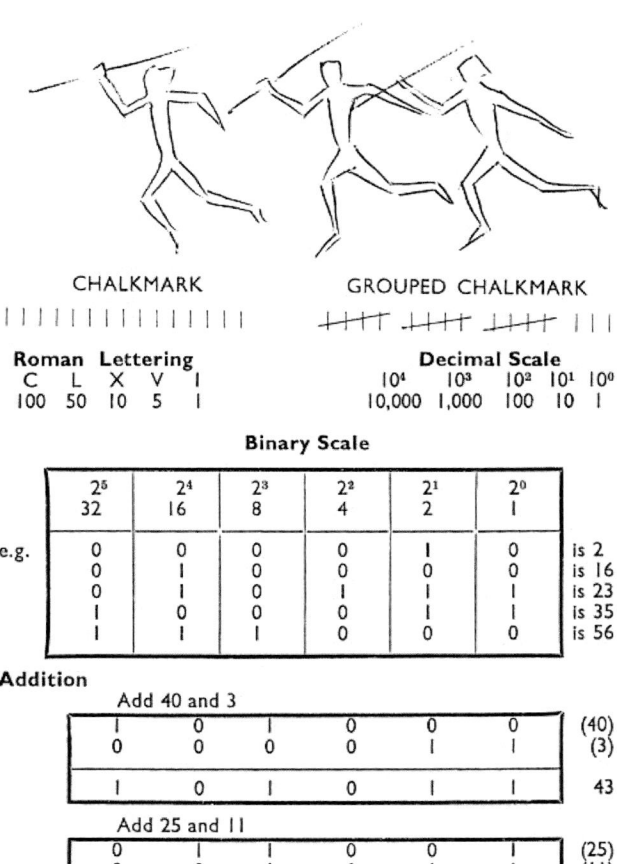

Fig. 4

way man has used numbers for thousands of years they are likely to be far less intimidating.

The earliest groups of numbers that we know of were representations on cave drawings (see Fig. 4). The great advantage of this was that one symbol only was used. The disadvantage, however, was that one symbol was used for each figure to be represented, and when large numbers were involved this became quite impracticable. Gradually these were vulgarised into grouped chalk marks. Eventually the Romans found a more practical way of packaging numbers, as is demonstrated in Fig. 4. The use of decimal numbers meant that enormous quantities could be represented by but a few symbols.

Gradually there was evolved the binary scale. The great advantage of this is that by locating just two numbers in the correct position any total number can be represented. Examples are provided, and the method of adding numbers is shown.

Since the basic characteristic of an electrical conductor is that it is either on or off, these two functions can be used to represent the two symbols in a binary code. Hence its widespread use in computer work.

SECTION 3
TECHNOLOGICAL CHANGE; ITS EFFECTS

As a trade union, we see our members not merely as units of production, which appear at a work-place each day, but as human beings in their full social and cultural context. It is therefore important to analyse the effects the equipment we have described is likely to have upon the "quality of life" of our members. Since TASS members are not an island unto themselves it will be necessary to analyse this in the broader context of the effect technological change is having upon the workforce as a whole. By doing this in a proper historical context it should be possible to identify discernible laws of development, and having understood these laws to use them to scientifically predict what effects the equipment will have on our members in the future.

CAD – Its Nature and Implications

In doing this, we shall inevitably concentrate on the dangers and contradictions which this kind of equipment in a profit-orientated society brings in its wake. If this appears negative it should be understood that it is merely done to counterbalance the verbal overkill of the salesmen who pose this high capital equipment as the solution to all our problems. We as a union are not opposed to technological change. Indeed, it is our members who design much of the equipment that makes technological change possible. We are however fundamentally opposed to the misuse of technology! We are not like some romantics, who seem to believe that before the industrial revolution the populace spent its time tripping through daisies in unspoiled meadows or dancing round maypoles.

We, as designers and technologists, are fully aware of the enormous contribution science and technology has made in eliminating disease, ending squalor and improving economically the quality of life. Our concern is the misuse of technology, which can amongst other things create a frantic work tempo for some and the dole queue for others. It is therefore the object of this Section to forewarn the trade union movement of some of these dangers.

Let us then briefly look at technological change from earliest times, in order to identify its laws of development. The earliest men we know of made and used tools in order to enable them to meet the primitive requirements of shelter, food and an economic environment in which to reproduce their own kind. The techniques by which they did this have been in constant change, and are generally referred to as technological change.

Technological development is a powerful force which has moulded the course of history from earliest times, not merely in the sense that it tends to raise the standard of living of all or sections of the community, but also in a much more profound political sense in that technological change alters the whole character of society.

It was the invention of agriculture and a subsequent flow of inventions such as metallurgy and the use of wheeled transport which transformed the simple life of primitive communism into

DEMOCRACY OF IRON

civilisation, with its complexities and class divisions.

About 3000 BC a discernible change was taking place in the structure of society. The communities of equal farmers were gradually replaced by states in which the vast majority lived at subsistence level, while all the surplus products of their labours were used for a small class of Kings, noblemen and religious leaders. Class division became the basis of social structure.

That age obtained its name from the metal used to produce the implements at that stage, bronze. However, due to its rarity and costliness, bronze never greatly extended man's control over nature. Its rarity also ensured that it was only available to the prosperous class. When man learned to produce iron as well as bronze the then society was profoundly affected by this technical advance. At this stage metal tools became generally available to the farmer, and enormously increased the productivity of agriculture. From 700 BC iron axes made possible the clearance of great forests, and hence a further expansion of agriculture. The increased productivity of agriculture yielded a surplus which could then support a large number of craftsmen. The commodities produced by the craftsmen became more generally available, and were no longer merely produced for the wealthy. The craftsmen provided the farmers directly with tools in order to increase the productivity of their work. There then existed for the first time a balanced relationship between industry and agriculture.

This changed relationship ended a stage in which agriculture provided the food for the craftsmen but the craftsmen's product went to the select few. The craftsmen, by using iron, were able to provide themselves with ever improved tools, thereby increasing the productivity of their craft, which in turn tended to enhance their economic status. Thus, the advance in technology from the production of bronze to that of iron tended to break down barriers between classes, which had brought about stagnation in the Bronze Age.

CAD – Its Nature and Implications

POWER DRIVEN MACHINERY

It can therefore be demonstrated that from the earliest times technological change has had a profound effect upon the structure of society. The more democratic Iron Age societies created circumstances in which technological advance could be made. This progress was, however, limited by the fact that although slavery created the conditions for the accumulation of wealth into fewer hands, and therefore laid the basis for the further development of productive forces, yet in its decline it was responsible for holding back full development of techniques such as animal power and the water wheel.

Thus, technological advance necessitated a social change, in which the slave states had to be replaced by mediaeval feudalism. This structure of society provided a higher status for the master craftsman, and thereby stimulated a wealth of technical innovation, including the first development of power driven machinery. By the end of the Middle Ages the scale and nature of machinery had become too large for the social organisation which created it. Thus, the master craftsmen and their powerful guilds which had introduced the machinery then became an impediment to further progress.

The further development of the productive forces could only be brought about by the newly arising capitalist class. Capitalism then provided the social organisation which made possible the primitive accumulation of capital, the social organisation for using heavy machinery, and the development of the economic framework within which it could be effectively deployed. Any reasonable history book will give graphic accounts of the appalling conditions imposed upon the working class to make this primitive accumulation of capital possible.

A RETARDING FORCE

We shall attempt to demonstrate that capitalism today, just as the societies which preceded it, has reached a historical stage when it is a retarding force, not only politically but in the field of technological change and in the implementation of the techniques which technological change now makes possible. Not least can this be identified in the field of computerisation. To generalise, one might say that technology and social organisation interact to elevate the nature of man's existence to a higher level. Each form of society during its early stages tends to encourage the advance of technology. However, when technological levels rise the society which has given birth to it tends to fail to keep pace with it. It then arrives at a historical breaking point at which the old society must be transformed into a new one, otherwise it is incapable of utilising the potential of technology to the full.

In capitalist society, science and technology, which could provide the material basis for a fuller and more dignified form of existence, is being retarded. Further, because of the enormous contradictions within capitalist society, technology is heightening those contradictions to levels where a change in society becomes even more imperative. Capitalist society is now incapable of providing the rational framework in which to organise the productive forces in social production.

RATE OF CHANGE

Before dealing with some of the laws of technological change and relating them to contemporary problems, it is necessary to quantify, if only roughly, the actual rate of this change. The scale of development in the last 20 years is probably equal to that accomplished in all of man's existence. The scale of scientific effort (which is closely related to technological change) in the present century has increased out of all recognition. It has been asserted by Professor J. D. Bernal, that in 1896 there were perhaps in the world some 50,000 people who between them carried on the whole tradition of science, not more than 15,000 of whom were responsible for the advancement of knowledge through research. Today, the total number of scientific workers in industry, government and academic circles is in the order of 2.5 million.

CAD – Its Nature and Implications

90% of all the scientists and technologists who ever lived in the whole of man's existence are alive today. Predictions of where this will lead are legion.

As the rate of technological change increases, so also does the rate at which knowledge becomes obsolete.

Sir Frederick Warner, in an article entitled 'Production Technology' pointed out that a chemical engineer has in fact made a mathematical model, representing this.

> "Let S represent the total stock of useful theoretical knowledge possessed by an engineer; F the fraction of this knowledge which becomes obsolete each year; R the fraction of the engineer's working time devoted to acquiring fresh theoretical knowledge; and L his learning rate. Then:
>
> $$LR = FS + \frac{ds}{dt} \quad (\frac{ds}{dt} \text{ is the rate of change of useful knowledge}).$$
>
> If it be assumed that (a) S is to remain constant and equal in amount to the stock of knowledge S_0 with which the engineer left the university; (b) five per cent of this knowledge becomes obsolete each year and (c) the man's rate of learning remains constant and equal to the average rate during his three year university course, then the equation becomes:
>
> $$R\frac{S}{3} = 0.05 S_o \text{ whence } R = 0.15$$
>
> Thus the engineer would need to spend 15% of his working time in order to keep his theoretical knowledge up to date; say a fortnight's course every year plus four or five hours a week devoted to studying books and magazines."

As might well be expected, the number of journals to be studied is ever increasing. This was shown by Hilary and Steven Rose in 'Science and Society'. In some fields, the rate of obsolescence of knowledge is even greater. This is particularly true in some areas of computer application. Norman Macrae, Deputy Editor of the

The Search for Alternatives

Economist states in the issue of January 22nd 1972:

> "The speed of technological advance has been so tremendous during the past decade, that the useful life of the knowledge of many of those trained to use computers has been about three years".

He further estimated that

> "A man who is successful enough to reach a fairly busy job at the age of 30, so busy that he cannot take sabbatical periods for study, is likely, by the age of 60 to have only about one eighth of the scientific (including business scientific) knowledge that he ought to have for proper functioning in his job".

It has even been suggested that if one divided knowledge into quartiles of outdatedness, those in the age bracket over 45 would find themselves in the same quartile as Pythagoras and Archimedes. These enormous rates of change are particularly evident in the design environment as it moves forward to computer aided design. The stress it places upon design staff, in particular upon older men, should not be under-estimated. The same kind of strain also applies to workers on the shop floor. This trend will continue as the computer spreads into other fields.

RATE OF OBSOLESCENCE

When we analyse the history of technological change it is clear that two major features are discernible in the type of equipment used for production. The first is that there is an ever increasing rate of obsolescence of this type of equipment. Steam engines made by Boulton and James Watt 200 years ago were still operating almost 100 years later. 100 years ago, when an employer purchased a piece of machinery, he could rest assured that it would last his lifetime and would be an asset he could pass on to his son. In the 1930s, machinery was obsolete in about 25 years; during the '50s it was ten years; at the moment, some machinery is obsolete in about five or six years.

Paul Chambers, when he was Chairman of ICI, spelt this out when he said that the company was:

> "beginning to think in terms of 15 years for new projects. In the early 1960s this came down to 12 to 15 years, and more recently the average for a new plant had come down to about ten years. For certain kinds of investment where the risks of technological obsolescence are thought to be high the amortisation period is down to five to seven years."

The second discernible feature is that the total amount of capital necessary to provide the means of production for a commodity on a mass production scale is ever increasing. A lathe, for example, 100 years ago would cost the equivalent of 10 to 20 men's wages per annum. A complicated lathe nowadays, with its NC tape control and the total environment necessary to prepare the tapes and operate the machine, will cost something in the order of 100 men's wages per annum. Thus, the cost of the means of production continues to rise, although that is not to say that the cost of commodities will rise.

Confronted with equipment which is getting obsolete literally by the minute, and has involved enormous capital investment, the employer will seek to recoup his investment by exploiting that equipment for 24 hours per day. In consequence of this, the employer will seek to eliminate all so-called non-productive time, such as tea breaks, will seek to subordinate the operators more and more to the machine in order to get the maximum performance, and will insist either that the equipment is worked on three shifts to attain a 24 hour exploitation, or is used on a continuous overtime basis. This trend has long since been self evident on the work shop floor. It is now beginning to be a discernible pattern in design areas.

High capital equipment, such as computer aided design equipment, is now becoming more widespread in technical areas. In consequence of that, the employers will wish to ensure that all employees who use this equipment accept the same kind of subordination to the machine that they have already established on

The Search for Alternatives

the work shop floor. To say that this is so is not to make a prediction about the far distant future. Many of our members will recall that during the Rolls Royce dispute of 18 months ago, which cost the union £1.25 million, the company sought to impose conditions of this kind at some of its plants. In Bristol, for example, they sought to get our members to accept a productivity agreement in which they demanded:

> "the acceptance of shift work in order to exploit high capital equipment, the acceptance of work measurement techniques, the division of work into basic elements, and the setting of times for these elements, such time spent to be compared with actual performance".

This harsh reality is very different from that envisaged by some Utopian Socialists who used to write books about the problems of people spending their leisure time when the work was automated and computerised. It is also in glaring contrast with those predictions that high capital equipment in the technological areas would simply liberate people from routine tasks and free them to devote themselves more fully to creative activities. In a profit orientated society automation and high capital equipment will only be introduced into narrow areas of the economy, there to be exploited to the maximum at 24 hours a day. The motive forces of capitalist society will prevent the widespread introduction of this kind of equipment and through its general use the introduction of a shorter working week, longer holidays and more leisure time. Indeed, the effect will be to create a frantic work tempo for some, whilst creating a permanent pool of unemployed persons.

More and more of our members who interface with this high capital equipment are being pressed to accept shift working. The effects of shift work on those who are compelled to undertake it are considerable in a physical, psychological and social sense. A number of studies demonstrate this. P. E. Mott in *Shift Work, the Social, Psychological and Physical Consequences*, Ann Arbor, 1965

> "found that day workers get an average of 7.5 hours sleep per night, which is one hour more than the overall average of rotating shift

CAD – Its Nature and Implications

workers, but when they are working the night segment of the shift rotating shift workers average only 5.5 hours of sleep. The biggest problem for rotating shift workers occurs when they move from their turn on the day shift to the night shift."

One study "reports that only 37 % of the workers adjust to the new sleeping times immediately, while 28 % of the workers said that they took four days to adjust to the night shift ... Another study of operators in two different plants in the United States found that only 31% of the men working under an extended 7-hour week rotation reported that they had adjusted to the harder shift change within a day or less. Even fewer, just 5% of the men, working a monthly rotation schedule, stated that they could adjust to the hardest shift change in one day. Under the latter schedule 70% reported that their adjustment to the new schedule took four days or more."

"A higher proportion of night and rotating shift workers reported that they were fatigued much of the time, that their appetites were dulled and that they were constipated much of the time".

"The ulcer rate amongst German workers was eight times as high for rotating shift workers as for the fixed shift group."

"The most frequently mentioned difficulties in husband/wife relationships concerned the absence of the worker from the home in the evenings, sexual relations, and difficulties encountered by the wife in carrying out her household duties."

"Another area of family life that seems to be adversely affected by certain kinds of shift work is the father/child relationship."

We can see therefore that the introduction of this kind of equipment in a profit orientated society can actually lower "the quality of life". The disruption of social life in consequence of shift working is also very great. I am well acquainted with a suburban estate in West London on which a number of mathematics graduates live. Some of them participated in a local amateur operatic society, others belonged to a small theatre group, some belonged to tennis clubs. All of these activities take place in the evenings. Recently, however, a large firm in the West London area, where some of them work, introduced a computerised system and it required these people to work on shift, with a consequential disruption of their social life. More importantly, as we have seen

The Search for Alternatives

from the demands of Rolls Royce, employers are attempting to subject their work to the same kind of detailed scrutiny that they have used for some considerable time on the shop floor.

When technical staff work in a highly synchronised computerised environment, the employer will seek to ensure that each element of their work is ready to feed in to the process at the precise time at which it is required. A mathematician will find that he has to have his work ready in the same way as a Ford worker has to have the wheel ready for the car as it passes on the production line.

A number of employers, in order to achieve this, have sought to introduce stop watch techniques. This has been vigorously resisted by our union, and we will continue to do so. In consequence of this, many graduates who in the past would never have recognised a need to belong to a real trade union, now find that they need the same kind of bargaining strength that manual workers have accepted on the shop floor for some considerable length of time. In fact, one can generalise and say that the more technological change and computerisation enters white collar areas the more the workers in those areas will become proletarianised. This proletarianisation will offer new recruitment opportunities for our union.

FRAGMENTATION OF SKILLS

Another major discernible feature of technological change is the fragmentation of skills. The millwright of 100 years ago was capable of repairing any machine in the plant in which he worked. He could predict the failure rate of bearings, select the material for new ones, and in most cases manufacture them himself. With the increasing complexity of machinery the jobs of the millwright were divided down into a series of specialised functions; firstly into maintenance, mechanical fitters and electrical fitters; gradually into machine tool fitters, prime mover maintenance fitters, and so the fragmentation continued.

Indeed now the failure rate of equipment is worked out by

maintenance planning and reliability engineers, sometimes using advanced mathematical techniques such as the theory of probability. In consequence, the division between manual and intellectual work becomes even greater. This fragmentation on the shop floor finds its parallel amongst technical white collar workers. The function of the design draughtsman is likewise becoming more and more fragmented and specialised. The draughtsman in the '30s was the centre of the design activity. He would design the component, draw it, stress it, select the materials for it, write the test specifications for it, liaise with the customer and usually liaise with the work-shop floor for production. Towards the end of the '30s, and certainly during the War, all of these functions were broken down into discernible separate jobs. The calculations were carried out by stressmen, the materials selected by metallurgists, the form of lubrication determined by tribologists, the draughtsman did the drawing, production engineering was carried out by methods and planning engineers, and customer liaison by specialist customer liaison engineers.

In order to exploit their high capital equipment to the full the employers will want our members to specialise more and more. Once they have achieved a degree of specialisation the employers will insist that they remain in that particular job whilst they are still needed. Fewer and fewer of our members will be able to see in a panoramic way the end product of the job on which they are involved. They will see only a tiny part of it. The effect will be that job satisfaction will be eliminated from more and more of our members.

It is now commonplace in industry to talk about dedicated machines and dedicated computers. That means dedicated to one specialised task. Since in a capitalist society man is seen as an appendage to the machine, that appendage will also become dedicated. The type of person required for a job will be specified precisely. Advertisements for engineering graduates now specify the precise requirements needed. No longer is a degree in engineering specified, but one in heavy engineering, light electrical engineering, electronics, electrical control, or one of a host of other

The Search for Alternatives

technical disciplines. It is in some of the fields of computer aided design that this over-specialisation finds its highest levels.

Indeed, the whole of the education system is now being geared to provide these highly specialised personnel. Less and less are people being educated to think; they are more and more being trained to carry out a specific function. It is clear from the response to lectures I have given at Universities and colleges of technology that the student movement is increasingly aware that it is now being prepared as industrial fodder for these large combines, and not being educated to think as civilised human beings.

The "Man Component"

Not only will men be suited to the productive roles in terms of their technical ability, but since they are looked upon by employers as units of production they will increasingly be considered as components within the total man/machine interface. Just as the life expectancy of a machine is now calculated, it is clear that a number of employers are beginning to do so in respect of their employees.

The developments on the shop floor should forewarn us of this.

Some agreements in high productivity plants seek to eliminate shop floor workers purely on the basis of age. In Standard Triumph in Coventry, it is reckoned that a man is burned up in ten years on the main production line. Clearly the employer wishes those years to be as early as possible, in order that the operator in question is active enough to exploit the plant at the highest rate possible. The company recently attempted to get some of the manual workers to agree that only workers of up to 30 years would be recruited for this high tempo work. Similar assessments are being made of the life span and peak activity age of intellectual workers. Recently, in the *Sunday Times*, a list was given of the peak performance age for mathematicians, engineers, physicists and others. For some of these the peak performance age was 29 or 30. A number of American studies have recently pointed out these peak performance ages, and suggested that this should be followed by a careers plateau for three or four years and thereafter, unless the employee in question has moved into management, that he be subjected to a "careers de-escalation."

Our own practical experience demonstrates, particularly during redundancies, that older men are being eliminated in this way. They are being eliminated, or down-graded to lower paid work, simply because they have committed the hideous crime of beginning to grow old. We are, as Samuel Beckett once said, "all born of the gravedigger's forceps". Growing old is the most natural process inherent in man. It is a biological process, but in the contradictory nature of the profit orientated society it is treated almost as a crime.

Much research has been carried out upon the effects of "ageing". Many of the experimental psychologists involved in this field of work, obviously hope that the findings of their work might result in suitable work forms being found for older men where their greater experience and more mature judgement might be offset against the slowing down which is the characteristic of ageing. The harsh reality of industrial life determines otherwise. K. F. H. Murrell, writing in *Ergonomics*, Volume 5 (1962) in a paper entitled 'Industrial aspects of ageing' (page 147-153) stated:

> "The purpose of ergonomics in this field then should be to reduce the initial demands of a job to a level at which all forseeable stresses will be within the capacity of men of all ages. **It is quite clear that industry is not thinking in these terms at present. Heron & Chown say 'when one turns to the idea of modifying jobs in favour of older men, we must report that not one manager in the 116 we interviewed had ever done so'**" (My emphasis – M.C.)

Yet the research continues and the findings are published. A. T. Welford in a paper entitled 'Changes on Speed of Performance with Age and their Industrial Significance' published in the Penguin Modern Psychology Readings series provided the results of some such experiments:

> "Several experiments have shown disproportionate slowing with age when complications are introduced into the relationships between

The Search for Alternatives

display and control. A simple example is when actions are guided by a display seen in a mirror as opposed to one seen directly (Szafran, see Welford, 1958). Perhaps the most striking demonstration, however, is in a set of five tasks studied by Kay (1954, 1955). All these used a box containing a row of twelve light bulbs and a corresponding box of twelve keys. The subject's task was to press the corresponding key as quickly as possible when a light came on. The arrangements of box and keys in the several tasks were as follows:

A. Each light was immediately above its corresponding key.

B. The same as A, but with the box of lights set 3ft. away across the table from the box of keys, so that in order to relate light to key the subjects had to align across a 3ft. gap.

C. The same as B, but with the lights related to the keys by means of a number code instead of by straight forward spatial correspondence. Subjects had to imagine the lights as numbered and to find the number of the light on a card bearing the numbers 1-12 in random order and placed immediately above the keys.

D. The same as C, but with the card placed mid-way between the lights and keys, so that subjects had both to use the number code and to align across a 1.5ft. gap.

E. The same as C, but with the card by the lights, so that both use of the number code and alignment across a 3ft. gap were required.

The disproportionate rise of time with age as the tasks became more difficult is very striking. The pattern of errors was similar so that there is no question of the older subjects having been slower because of having been more accurate.

Two points should be noted. Firstly, the times include time taken to correct errors and are of a quite different order from those of reaction time experiments; and secondly, difficulty in both types of task appeared to depend upon failure with age in some form of short term retention."

We shall have to ensure that this form of research is not used to identify, on the basis of age, those to be subjected to a "careers de-escalation" into the dole queue.

All our younger members should fully understand that however energetic and forceful they may feel now they will all begin to grow old, and if they allow our older members to be treated in this way they are creating a framework of oppression which will be

CAD – Its Nature and Implications

used against them in the future.

We, as a union, must assert that those who over the years have created the real wealth which makes possible the purchase of high capital equipment should have the right of working at a civilised tempo in the autumn of their lives.

Many employers use the introduction of the kind of equipment described earlier in the booklet as a smoke screen under which to get rid of older men. They imply that the techniques are so involved, complicated and new that only up and coming technologists can cope with them. It is essential that we guard against these tendencies.

The introduction of computerised systems is also used on occasions as a basis for the introduction of that great pseudo-science, job evaluation. So called scientific reasons are given to justify dividing our members into a hierarchical structure. In the process many jobs are fragmented, and the fragmented component parts are given very low levels of payments. In some instances, employers seek to suggest that they are actually semi-clerical. Our experience shows that a number of companies attempt to use this fragmentation to further consolidate the unequal pay between men and women. They seek, through these job evaluation schemes, to assert that certain fragmented functions are women's work. There is no such thing as women's work, any more than there is women's mathematics, women's science, women's literature or women's music. There is only work, and we should continue to demand that all our members in a department are paid decent wages for undertaking it irrespective of sex or age!

When introducing this kind of high capital equipment, employers will seek to ensure when they analyse the man/machine system that the response rate of the man is quick enough to optimise the performance of the equipment. The experience of shop floor workers will again help to identify these trends. More and more productivity deals are being drawn up in which an assessment is made of the man's capabilities. In some of these this is even done through medical examination. In a civilised society a medical examination would simply mean that if something were

The Search for Alternatives

the matter with you, you could go to a hospital, have appropriate treatment, and return to your job.

In the productivity agreement reached in the steel industry, and known as the Green Book, such medical examinations are an inherent part. It would appear, however, that these medical checks are to examine the response rate of the worker and his ability to interface with high tempo equipment. In some instances, where a worker's health has been impaired due to the intense rate of work, the medical check is used as a means of down-grading him. In steel some workers have lost up to £15 a week in consequence of such medical examinations.

This down-grading finds its analogy in the machine component when such a machine, having been involved in high precision work for a number of years, deteriorates and is then relegated to second-rate work in the jobbing shop.

This systematising of people in the same way as units of production further subjugates them to the machine, and creates enormous pressures for them. The same thing is being done in a more subtle manner to white collar workers. In many instances the techniques are introduced in the guise of management development programmes, or suitability assessment schemes. Some of them in the United States seek to assess that the dynamic rate of thinking of the technical worker is fast enough for him to respond to the computerised systems he is operating. Once again, it is the older man who is driven to the wall.

INTENSIFICATION OF WORK RATE

The actual rate at which design staffs operate will be intensified by the use of computerised equipment. The Department of Labour in the United States on one occasion assessed that a designer doing a job in the aerospace industry spent 95% of his time undertaking reference work and only 5% on actual design decision making. The introduction of computer graphic systems, described elsewhere, can eliminate the routine reference work and actually intensify the decision making rate by up to 1900%. The stress this

CAD – Its Nature and Implications

will put upon the design staff involved can be enormous.

When one looks at the total mathematical system, the "man component" is slow, inconsistent, unreliable, but highly creative, whereas the computer is fast, consistent, and reliable, but totally non-creative. When the two are interfaced there is a perfect unity of opposites, but the load on the man as he attempts to respond to the speed of the system is very great indeed. It will be essential for our union to ensure that men subjected to this kind of pressure are provided with proper rest periods. Those involved in working interactively with the computer will find that the system continuously compels them "to increase their dynamic rate of thinking, frequency response or phase lag."

The "man-component's" response rate is measured in the same dehumanised way as a diode might be. In fact man is seen merely as a component in a servo mechanism.

As far back as 1960 E. R. F. W. Crossman was saying in the *Quarterly Journal of Experimental Psychology*:

"The usual experimental pursuit-tracking task consists of a display with a pointer which moves from side to side providing a course or input, $i(t)$. The subject has a control connected to a controlled member, a pointer or pen, which produces a track or output, $o(t)$, and his task is to keep the controlled member aligned with the course pointer so that the error, $e(t)$, is a minimum where

$$e(t) = o(t) - i(t)$$

In engineering terms the subject is then part of a closed-loop control system or servo-mechanism whose function is to 'follow-up' the coursepointer. There is a well developed mathematical theory of linear servomechanisms and one of its central concepts is the Transfer function, which gives $o(t)$ as a function of $e(t)$ and its differential co-efficients; hence several workers have attempted to specify the human transfer function and they have had some success (Hick & Bates, 1951; North, 1950). Others (e.g. Fitts, Noble and Warren, 1955) thinking along similar lines have measured subjects 'frequency response' and 'phase lag' when following sinusoidal inputs."

The Search for Alternatives

The rate at which they will be required to make decisions continues to increase all the time. In the past, the freedom to walk about to a library to gain reference material was almost a therapeutic necessity. The opportunities to discuss design problems with one's colleagues often resulted in a useful cross-fertilisation of ideas, and in a resultant better design. As more and more interactive systems are evolved and software packages built up for them, man's knowledge will be absorbed from him at an ever increasing rate, and stored in the system. In consequence, he will converse more with the machine and less with other people. All this will mean that the design environment will become more capital intensive as the organic composition of capital changes and people are replaced by machines. This is one contributory factor to the growing unemployment as many jobs will be permanently eliminated and unemployment will begin to grow. Nor is this peculiar to the UK. It is already evident in some of the monopolies in this country that they are concentrating their design power in specialised computer centres, whilst at the same time they are declaring design staff redundant elsewhere. The staff engaged at the design centres are being compelled to work either consistent overtime or on shift, whilst their colleagues have to join the dole queue. Our members must therefore ensure that this kind of equipment is only introduced in such a phased fashion as to ensure that it meets actual work load requirements, and is not used to eliminate members at other sites. It should, however, always be borne in mind that this kind of equipment is seldom introduced to effect savings on wages. In very few companies, even in the United States, in spite of savings of 8:1 on actual man hours, is a saving effected on wages. The high capital equipment is so costly that it actually off-sets the saving in wages.

The real driving force to introduce this kind of equipment is in order to reduce the "lead time". This became more and more a problem for large companies, as millions of pounds worth of equipment is tied up in castings and unfinished products on the shop floor, so the equipment tends to be introduced in order to reduce the time scale between the original design concept and the sale of the end product.

Consequences for Manual Workers

The use of some of the equipment described changes the relationship of the productive forces in large organisations. In many instances the equipment is used not only at the design stage but to produce the manufacturing data and NC tapes for machines on the shop floor. A number of companies now use automatic draughting equipment, described above.

We pointed out that the draughtsman uses a digitiser to produce a tape and not a drawing. We showed that the tape could be used to create a drawing on a micro-plotter, but more importantly it may be used directly on a numerically controlled jig bore, lathe, milling machine or other machine tool. This means in practice that some of the most highly skilled jobs on the shop floor are permanently eliminated and the machine setting function is transferred to white collar workers in the drawing office, whilst the operation of the machines is handed over to semi-skilled workers.

The effect this can have upon the Engineering Section of our union is enormous.

It is therefore imperative that when this kind of equipment is being introduced our members should discuss it fully with the shop stewards' committee and only agree to its introduction when they are satisfied that their members' interests on the shop floor have been safeguarded. Not only are some of the most highly skilled machine setting and operating jobs, such as that of a jig borer being eliminated but other skilled functions are also affected. The same tape, in the system described, can be actually used to inspect components on a digital inspection machine. In consequence of this the inspection function can be de-skilled. Even on those machines where the operator still requires very considerable skill, the effect of introducing numerical controlled equipment is that it is possible to produce a fully trained operator in a very short length of time.

At Alfa Laval in Sweden the turners on the big vertical boring

The Search for Alternatives

mills, turning the outer casings of centrifugers, used to take seven years to train and gain the necessary experience. The company now finds that by controlling this machine numerically they can produce a fully skilled operator in one year. Again, the impact this will have on our Engineering Section will be very great indeed.

In some industries the tool making function will be largely eliminated by the use of spark erosion machines to generate the tools. It is significant that the company, in introducing this kind of equipment, pointed out that the resistance of tool makers to accepting these techniques can be eliminated by getting them to do their own programming. They are therefore in a situation of preparing the rods for their own backs, unless they get very firm guarantees indeed!

The consequences for semi-skilled workers can be just as great. In those cases where there is still a high labour content in the production process, the tape can also be used to control them. The effect in this situation is that the operator completely loses control over the tempo at which he works. The tape sets the tempo and the operator must either respond to it or stop the machine. In some systems every stoppage of the machine is recorded by a central computer. The operator is thereby ensnared in a system which increasingly subordinates him to the machine.

The contradictions which arise cannot be resolved within the framework of a free enterprise system, since they are but manifestations of the irreconcilable contradictions between the interests of the exploiter and the exploited. Trade unions, however militant, can at the best only protect their members from the worst excesses of technological change in this form. Only a political solution, a change in the ownership of the means of production, can harness these new forces in the interests of the majority of the population.

In the meantime, the very technological advances which could liberate man from drudgery and provide the material basis for a full and prosperous life will, under this system, bring about the material and spiritual impoverishment of the people.

CAD – Its Nature and Implications

INCREASED STRIKE POWER

We have concentrated up to now in demonstrating some of the negative effects this kind of equipment can have upon our members and the rest of the trade union movement. The equipment brings in its wake a host of problems for the employer, other than that of the increased rate of obsolescence and high capital investment already described.

One of the contradictions for the employer is that the more capital he accumulates in any one place the more vulnerable it becomes. The more closely he synchronises his industry and production by using computers the greater becomes the strike power of those employed in it. Mao Tse Tung once said, in his military writings, that the more capitalised an army becomes the more vulnerable it becomes also. This has been demonstrated in Vietnam, where a Vietcong with a 30/- shell can destroy an American aircraft with airborne computer costing something like £2.5 million. The capitalisation of industry also produces an analagous situation. In the past, when a clerical worker went on strike it had precious little effect. Now if the wages of a factory are carried out by a computer a strike by clerical workers can disrupt the whole of the plant.

It is also true on the factory floor that in the highly synchronised motor car industry a strike of 12 workers in the foundry can stop large sections of the entire motor car industry.

The same is happening in the design area. As high capital equipment, through computer aided design, is being made available to design staffs, firstly it proletarianises them, but secondly it also increases their strike power. In the past when a draughtsman went on strike he simply put down his 6H pencil and his rubber, and there was unfortunately a considerable length of time before an effect was felt upon production, even when the manual workers were blacking his drawings. With the new kind of equipment described, where NC tapes are being prepared or where high capital equipment is used for interactive work, the effects of a strike will in many instances be immediate, and production will be affected in a very short length of time. This greatly increases the strike power of our members, and will put up their bargaining power considerably.

The employers, however, realising that this is so and that more and more white collar workers are looking for effective trade union organisation, will seek to designate these areas of management in order to prevent them becoming effectively organised. They will seek either to introduce into them tame trade unions or anti-union collaborationist organisations such as UKAPE (UK Association of Engineers, founded in 1969). The danger in this is twofold. Firstly, it will prevent the growth of real trade unionism in these areas, and worse still it could provide an anti-trade union anti-working class base in a very important controlling sector of industry, which could be used against manual and other workers on the shop floor.

Recognising that this is so, and recognising the contradictions that technological change introduce for white collar workers, we should be able to expand rapidly in these areas. Computerisation will also extend TASS's power in another sense. We have demonstrated that much of the design know-how is contained in software packages.

More and more of these companies are now being set up throughout the country as a design service to industry. Since these people are selling design know-how they should be treated by our members in exactly the same way as contract drawing offices were. With the power our members have in the parent companies and in the manufacturing units, they should be able to use that strength to compel these sectors to become organised.

Man, "the Machine Appendage"

Throughout this section dealing with the effects of technological change a number of references have been made to employers regarding man as an appendage to the machine. To some readers this may seem an exaggerated or a polarised analogy. Let us therefore look objectively at how a unit of production is designed,

CAD – Its Nature and Implications

and then analyse the concrete situation in which we find ourselves as human beings within this system and make a realistic comparison.

We know, as designers, that when you design a unit of production you ensure that you design it to operate in the minimum environment necessary for it to do its job. You seek to ensure that it does not require any special temperature controlled room unless it is absolutely essential. In designing the lubrication system you do not specify any exotic oils as lubricants unless it is necessary. You ensure that its control system is provided with the minimum brain necessary for it to do its job. You don't, for example, have a machine tape controlled if you can get away with a manual one. Finally, you provide it with the minimum amount of maintenance, in other words, you design for it the maximum life span in which it will operate before a failure.

Those who control our society see human beings in the same way. The minimum environment for the man means that you provide him with the absolutely lowest level of housing which will keep him in a healthy enough state to do his job. If one doubts that, it is still worth remembering that 7,000,000 people live in slums in Britain. The equivalent of fuel and lubrication for the machine is the food provided for a man. This is also kept at a minimum for those who work. We even find Oxford dieticians telling old age pensioners how they can manage on £2 of food per week. The minimum brain is provided for the man by an educational system which gives him enough knowledge to do his job, which trains him to do his job, but does not educate him to think about his predicament or that of society as a whole. The minimum maintenace necessary is provided through the National Health Service, which concentrates on curative medicine rather than preventive medicine, and the reality, the harsh reality, is that when a man has finished his working life he is thrown on the scrap heap like an obsolete machine.

If that is felt to be an extreme position it is worth recalling the statement of the doctor at Willesden Hospital, who said there was no need for National Health patients over the age of 65 to be resuscitated. This is the harsh reality of the manner in which employers regard us. If anybody doubts that that is so they will soon find out if they go blind or become seriously ill.

CONCLUSION

One is frequently asked if technological change is a good thing or not. That is really a non-question. It depends entirely how it is used and who controls it. We need not have any fear of technological change. It is in fact merely an extension of man's own capabilities. Historically man sought to extend his eyes by using telescopes, ranging from the time of Galileo to today where he uses radar and radio telescopes. He extended his tongue and his ears by communication systems and audio aids. His muscular power he extended through mechanisation, and his energy he extended and increased through the harnessing of nuclear power. Even the most sensitive faculty of man, that of memory and his nervous system, he has now in many ways extended by the decision-making techniques used through computers. It was us and people like us who used their great skill and ingenuity to create all this technological change.

Our members long to be able to use that skill and ingenuity to provide the material basis for a more full and dignified existence for the community as a whole. This drive for further scientific knowledge "into that untravelled world whose margin fades forever and forever when I move" is to be welcomed – indeed it is one of the guarantees of our future prosperity. It must not however be a blind unthinking drive forward. As the main Union whose members design much of the equipment described in this booklet we have a social and political responsibility to analyse its effects upon our members and the class to which they belong in our profit orientated society. If this booklet provides the stimulus for such an analysis its purpose will have been well served.

CAD – Its Nature and Implications

OUR UNION POLICY ON COMPUTER AIDED DESIGN

1. Insist on the fullest consultation with companies proposing to introduce computer systems.
2. Organise the whole staff on such installations.
3. Organise establishments producing programmes and software packages for design work in the main company in the same way as we organize contract drawing offices.
4. Protect the employment of existing personnel including insisting on retraining to staff the new process.
5. Oppose redundancy on the basis that such equipment should only be introduced to fulfil an actual existing work load requirement.
6. Resist work measurement techniques and other methods used to try and intensify the work rate.
7. Insist on realistic wage levels.
8. Press for the 35 hour week, longer holidays and proper rest periods to be brought in when computer systems are introduced as a counter to the increase in stress and work tempo.
9. Oppose extensions of shift-working.
10. Keep the union advised.

EDITORS NOTE
Sections 1 and 3 of Computer Aided Design – Its Nature and Implications *are published here. Section 2, a technical and mathematical exposition of the scene in the 1970s, has been omitted.*

Contradictions of Science and Techology in the Productive Process

First published in *The Political Economy of Science*, 1976

Any meaningful analysis of scientific abuse must probe the very nature of the scientific process itself, and the objective role of science within the ideological framework of a given society. As such, it ceases to be merely a 'problem of science' and takes on a political dimension. It extends beyond the important, but limited, introverted soul-searching of the scientific community, and recognises the need for wider public involvement. Many 'progressive' scientists now realise that this is so, but still see their role as the interpreters of the mystical world of science for a largely ignorant mass, which when enlightened will then support the scientists in their intention 'not to use my scientific knowledge or status to promote practices which I consider dangerous'.

Those who, in addition to being 'progressive' have political acumen, know that a 'Lysistrata movement', even if it could be organised, is unlikely to terrify monopoly capitalism into applying science in a socially responsible manner. 'Socially responsible' science is only conceivable in a politically responsible society. That must mean changing the one in which we now live.

One of the prerequisites for such political change is the rejection of the present basis of our society by a substantial number of its members, and a conscious political force to articulate that contradiction as part of a critique of society as a whole. The inevitable misuse of science, and its consequent impact upon the lives of an ever-growing mass of people, provides the fertile

The Search for Alternatives

ground for such a political development. It should constitute an important weapon in the political software of any conscious revolutionary.

Even Marxist scientists seem to reflect the internal political incestuousness of the scientific community, and demonstrate in practice a reluctance to raise these issues in the mass movement. Thus the debate has tended to be confined to the rarified atmosphere of the campus, the elitism of the learned body or the relative monastic quiet of the laboratory.

Clearly, the ruling class, which has never harboured any illusions about the ideological neutrality of science, will not be over-concerned by this responsible disquiet. The Geneens of ITT and the Weinstocks of GEC do not tremble at the pronouncements of Nobel Laureates. It is true of course that the verbal overkill of the ecologist has reverberated through the quality press and caused some concern – not all of it healthy – in liberal circles. But the working class, those who have it within their power to transform society, those for whom such a transformation is an objective necessity, have not as yet been really involved. Yet their day-to-day experience at the point of production brutally demonstrates that a society which strives for profit maximisation is incapable of providing a rational social framework for technology (which they see as applied science).

'Socially irresponsible' science not only pollutes our rivers, air and soil, provides CS gas for Northern Ireland, produces defoliants for Vietnam and stroboscopic torture devices for police states. It also degrades, both mentally and physically, those at the point of production, as the objectivisation of their labour reduces them to mere machine appendages. The financial anaesthetic of the 'high-wage (a lie in any case) high-productivity low-cost economy' has demonstrably failed to numb workers' minds to the human costs of the fragmented dehumanised tasks of the production line.

There are growing manifestations in the productive superstructure of the irreconcilable contradictions at the economic base. The sabotage of products on the robot-assisted line at General Motors Lordstown plant in the United States, the 8 per cent

Contradictions of Science and Technology

absentee rate at Fiat in Italy, the 'quality' strike at Chryslers in Britain and the protected workshops in Sweden reveal but the tip of a great international iceberg of seething industrial discontent. That discontent, if properly handled, can be elevated from its essentially defensive, negative stance into a positive political challenge to the system as a whole.

The objective circumstances for such a challenge are developing rapidly as the crushing reality is hammered home by the concrete experience of more and more workers in high capital, technologically based, automated or computerised plants. In consequence, there is a gradual realisation by both manual and staff workers that the more elaborate and scientific the equipment they design and build, the more they themselves become subordinated to it, that is to the objects of their own labour. This process can only be understood when seen in the historical and economic context of technological change as a whole.

SCIENCE AND THE CHANGING MODE OF PRODUCTION

The use of fixed capital, that is, machinery and, latterly, computers, in the productive process marked a fundamental change in the mode of production. It cannot be viewed merely as an increase in the rate at which tools are used to act on raw material. The hand tool was entirely animated by the workers, and the rate at which the commodity was produced – and the quality of it – depended (apart from the raw materials, market forces and supervision) on the strength, tenacity, dexterity and ingenuity of the worker. With fixed capital, that is the machine, it is quite the contrary in that the method of work is transformed as regards its use value (material existence) into that form most suitable for fixed capital. The scientific knowledge which predetermines the speeds and feeds of the machine, and the sequential movements of its inanimate parts, the mathematics used in compiling the numerical control programme, do not exist in the consciousness of the operator; they are external and act through the machine as an alien force. Thus science, as it manifests itself to the workers through fixed capital, although it is merely the accumulation of the knowledge and skill now appropriated, confronts them as an alien and hostile force, and further subordinates them to the machine. The nature of their activity, the movements of their limbs, the rate and sequence of those movements – all these are determined in quite minute detail by the 'scientific' requirements of fixed capital. Thus objectivised labour in the form of fixed capital emerges in the productive process as a dominating force opposed to living labour. We shall see subsequently when we examine concrete situations at the point of production that fixed capital represents not only the appropriation of *living* labour, but in its sophisticated forms (computer hardware and software) appropriates the scientific and intellectual output of the white-collar workers whose own intellects oppose them also as an alien force.

The more therefore that workers put into the object of their labour, the less there remains of themselves. The welder at General Motors who takes a robotic welding device and guides its probes through the welding procedures of a car body is on the one hand building skill into the machine, and deskilling themselves on the other. The accumulation of years of welding experience is absorbed by the robot's self-programming systems and will never be forgotten. Similarly, mathematicians working as stressmen in an aircraft company may design a software package for the stress analysis of airframe structures and suffer the same consequences in their jobs. In each case they have given part of themselves to the machine and in doing so have conferred life on the object of their labour – but now this life no longer belongs to them but to the owner of the object.

Since the product of their labour does not belong to the workers, but to the owner of the means of production in whose service the work is done and for whose benefit the product of labour is produced, it necessarily follows that the object of the workers' labour confronts them as an alien and hostile force, since it is used in the interests of the owner of the means of production. Thus this 'loss of self' of the worker is but a manifestation of the fundamental contradictions at the economic base of our society. It

Contradictions of Science and Technology

is a reflection of the antagonistic contradiction between the interest of capital and labour, between the exploiter and the exploited. Fixed capital, therefore, at this historical stage, is the embodiment of a contradiction, namely that the means which could make possible the liberation of the workers from routine, soul-destroying, back-breaking tasks, is simultaneously the means of their own enslavement.

It is therefore obvious that the major contradiction can only be resolved when a change in the ownership of the means of production takes place. Much less obvious, however, is whether there exists a contradiction (non-antagonistic) between science and technology in their present form and the very essence of humanity. It is quite conceivable that our scientific methodology, and in particular our design methodology, has been distorted by the social forces that give rise to its development. The question therefore must arise whether the problems of scientific development and technological change, which are *primarily* due to the nature of our class-divided society, can be solved solely by changing the economic base of that society.

The question is not of merely theoretical interest. It must be a burning issue in the minds of those in Vietnam who are responsible for their country's programme of reconstruction. It must be of political concern to those in China, to establish if Western technology can be simply applied to a socialist society. Technology at this historical stage, in a class-divided society, such as Britain, is the embodiment of two opposites – the possibility of freeing the workers, yet the actuality of ensnaring them. The possibility can only become actuality when the workers own the object of their labour. Because the nature of this contradiction has not been understood, there have been the traditional polarised views: 'technology is good'; and 'technology is bad'. These polarised views are of long standing and not merely products of space-age technology. From the earliest times a view has persisted that the introduction of mechanisation and automated processes would automatically free people to engage in creative work. This view has persisted as consistently in the field of intellectual work

The Search for Alternatives

as it has in that of manual labour. As far back as 1624, when Pascal introduced his first mechanical calculating device he said,

> 'I submit to the public a small machine of my own invention, by means of which you alone may without any effort perform all the operations of arithmetic and may be relieved of the work which has so often fatigued your spirit when you have worked with the counters and with the pen.'

Only twenty-eight years earlier in 1596 an opposite view was dramatically demonstrated when the city council of Danzig hired an assassin to strangle the inventor of a labour-saving ribbon-loom, a defensive if understandable attempt, repeated time and again in various guises during the ensuing 500 years to resolve a contradiction at an industrial level when only a revolutionary political one would suffice. It is of course true that the contradiction manifests itself in industrial forms even to this day.

THE OBSOLESCENCE OF FIXED CAPITAL

There is first the ever shorter life of fixed capital (the increasing rate of obsolescence of machinery). Early wheeled transport existed in that form for thousands of years; steam-engines made by Boulton and Watt two hundred years ago were still operating about a hundred years later; a century ago, when an employer purchased a piece of machinery, he could rest assured that it would last his lifetime and would be an asset he could pass on to his son.

In the 1930s machinery was obsolete in about twenty-five years, during the 1950s in ten years, and at the moment computerised equipment is obsolete in about three to five years. Then there is the growing volume of fixed capital necessary to provide the total productive environment for a given commodity – the cost of the total means of production is ever-increasing. That is not to say that the cost of individual commodities will continue to increase. The most complicated lathe one could get 100 years ago would have cost the equivalent of ten workers' wages per annum. Today, a

Contradictions of Science and Technology

lathe of comparable complexity, with its computer-tape control and the total environment necessary for the preparation of those tapes and the operation of the machine, will cost something in the order of a hundred workers' wages per annum. The industrial manifestations of the contradiction now begin to emerge very clearly indeed. Confronted with equipment which is getting obsolete literally by the minute, and has involved enormous capital investment, the employer will seek to recoup his investment by exploiting that equipment for twenty-four hours per day. In consequence of this, employers will seek to eliminate all so-called non-productive time, such as tea breaks, will seek to subordinate the employees more and more to the machine in order to get the maximum performance, and will insist that the equipment is either worked upon on three shifts to attain a twenty-four hour exploitation, or is used on a continuous overtime basis. This trend has long since been evident in the manual field on the workshop floor. It is now beginning to be a discernible pattern in a whole range of white-collar occupations.

THE PROLETARIANISATION OF INTELLECTUAL WORKERS

An analysis of this problem in British companies demonstrates that employers will wish to ensure that all their white-collar employees who use this kind of equipment accept the same kind of subordination to the machine that they have already established for manual workers on the shop floor. To say that this is so is not to make a prediction about the far-distant future. In 1971 my union (AUEW-TASS) was involved in a major dispute with Rolls Royce, which cost the union £250,000. The company sought, amongst other things, to impose on the design staff at the Bristol plant the following conditions:

> "The acceptance of shift work in order to exploit high capital equipment, the acceptance of work measurement techniques, the division of work into basic elements, and the setting of times for these elements, such time to be compared with actual performance."

The Search for Alternatives

In this instance the union was able, by industrial action, to prevent the company from imposing these conditions. They are, however, the sort of conditions which employers will seek increasingly to impose upon the white-collar workers. When staff workers, whether they be technical, administrative or clerical, work in a highly synchronised, computerised environment, the employer will seek to ensure that each element of their work is ready to feed into the process at the precise time at which it is required. Mathematicians, for example, will find that they have to have their work ready in the same way as a Ford worker has to have the wheel ready for the car as it passes him on the production line. In consequence of this, many graduates, who in the past would never have recognised the need to belong to a real trade union, now find that they need the same kind of bargaining strength that manual workers have accepted on the shop floor for some considerable length of time. In fact, one can generalise and say that the more technological change and computerisation enters white-collar areas, the more workers in those areas will become proletarianised. The consequence of this will not be limited to the work situation. They will spread right across the family, social and cultural life of the white-collar worker. Consider the consequences of shift working for example. In a survey carried out in West Germany it was demonstrated that the ulcer rate amongst those working a rotating shift was eight times higher than amongst other workers. Other surveys have shown that the divorce rate amongst shift workers is approximately 50 per cent higher than normal, whilst the juvenile delinquency rate of their children can often be 80 per cent higher. There are a whole series of examples in Britain of the manner in which the cultural and social life of AUEW-TASS members has been disrupted by the introduction of this kind of equipment.

Thus, whilst it is true that automated and computerised equipment *could* free people from routine, soul-destroying, back-breaking tasks, and free them to engage in more creative work, the reality in our profit orientated society is that in many instances it actually lowers 'the quality of life'.

Contradictions of Science and Technology

There are also good grounds for assuming that automated and computerised systems will in many instances diminish rather than enhance the creativity of scientific and technological workers. Computer Aided Design (CAD) is a useful occupational aperture through which to view a scenario that will become commonplace to many in the next few years.

In selling the idea of computers to the design community, it is suggested that the computer will merely deal with the quantitative factors and the designer will deal with value judgements and the creative elements of the design process. It is of course true that the design process is, amongst other things, an interaction of the quantitative and the qualitative. It is not, however, true that design methodology is such that these can be separated into two disconnected elements which can then be applied almost as chemical compound. The process by which these two opposites are united by the designer to produce a new whole is a complex and as yet ill-defined and ill-researched area.

The sequential basis on which the elements interact is of extreme importance. The nature of that sequential interaction, and indeed the ratio of the quantitative to the qualitative, depends on the commodity under design consideration. Even where an attempt is made to define the proportion of the work that is creative, and the proportion that is non-creative, what cannot be readily stated is the stage at which the creative element has to be introduced when a certain stage of the non-creative work has been completed. The very subtle process by which designers review the quantitative information they have assembled, and then make the qualitative judgement, is extremely complex. Those who seek to introduce computerised equipment into this interaction attempt to suggest that the quantitative and the qualitative can arbitrarily be divided, so that the computer handles the quantitative. This is in reality a devious introduction of 'Taylorism' into advanced technological work – an attempt to further subdivide an 'intellectual activity' into its 'manual' and 'intellectual' components.

Since CAD dramatically increases the rate at which the quantitative is handled, a serious distortion of this dialectical

The Search for Alternatives

interaction takes place, frequently to the detriment of the qualitative. There are therefore good grounds for assuming that the crude introduction of the computer into the design process, in keeping with the Western ethic of 'the faster the better', may well result in a deterioration of the design quality. It is typical of the narrow, fragmented and short-term view which capitalism takes of all productive processes, that these important philosophical considerations are ignored. Much design research is limited to considerations of design techniques and associated hardware and software with precious little regard for the objective requirements of the design staff or, more importantly, the public. Such research accurately reflects our economic base: equipment and hence capital first; people last.

Elitist designers, steeped in their traditional professionalism, believed (and many still believe!) that their creative talents provide an eternal occupational sanctuary against the creeping proletarianisation of all white-collar workers. Architects, for example, conceded that there might be problems for aircraft designers, mechanical designers or civil engineers, but not for them. After all, is not architecture the 'Queen of the Arts rather than the Father of Technology'? However, capitalism, in its relentless drive to exploit all who work, has not forgotten the architects. For them there has been specifically produced a software package known (appropriately) as HARNESS. The concept behind this system is that the design of buildings can be systematised to such an extent that each building is regarded as a communication route. Stored within the computer system are a number of predetermined architectural elements which can be disposed around the communication route on a Visual Display Unit to produce different building configurations. Only these predetermined elements may be used and architects are reduced to operating a sophisticated 'Lego' set. Their creativity is limited to choosing how the elements will be disposed rather than considering in a panoramic way the types and forms of elements which might be used. As Marx pointed out in *Capital:*

Contradictions of Science and Technology

'a bee puts to shame many an architect in the construction of her cells. But what distinguishes the worst architect from the best of bees is this, the architect raises his structure in imagination before he erects it in reality. At the end of every labour process, we get a result that already existed in the imagination of the labourer at its commencement'.[1]

It is clear that HARNESS will reduce the distinction between the architect and the bee and that capitalism will insist that in future architects will work in a more 'bee-like' fashion! This will gradually apply to all technological and scientific workers, whatever systems are devised to control their 'mode of intellectual production' in the same way as manual workers on the shop floor are controlled. Employers have long sought ways and means of controlling 'their' elusive, creative technical and white-collar staff. The computerised system is one Trojan Horse widely used to do so. The process is succinctly described in the magazine *Realtime* by a writer fresh from an IBM customer training course:

"Now an operating system is a piece of software, functionally designed to do most efficiently a particular job – or is it? It gradually dawned on me that some rather obnoxious cultural assumptions have been imported lock, stock and barrel into IBM software. Insidious, persuasive assumptions, which appear to be a natural product of logic – but are they? The whole thing is a complete totalitarian hierarchy. The operating system runs the computer installation. The chief and most privileged element is the 'Supervisor'. Always resident in the most senior position in main storage, it controls, through its minions, the entire operation. Subservient to the supervisor is the bureaucratic machinery – job management routines, task management, input/output scheduling, space management and so on. The whole thing is thought out as a rigidly controlled, centralised hierarchy. And as machines get bigger and more powerful, so the operating system grows and takes more powers. One lecturer soared into eloquence in comparing the various parts of the operating system to the directors, top management, middle management, shop foremen and ordinary pleb workers of a typical commercial company. In fact, the whole of IBM terminology is riddled with class expressions – master files, slave cylinders, high and low level languages, controller, scheduler, monitor."[2]

The Search for Alternatives

The same writer then generalised some of the contradictions of centralised operating systems. These coincided closely with my own findings when I investigated the contradiction in the specific field of Computer Aided Design.

> "The drawbacks of the centralised operating system are many. It is a constraining and conservative force. A set of possibilities for the computer system is chosen at a point in time, and a change involves regeneration of the system. It imposes conformity on programming methods and thought. Another amazingly apt quote from an IBM lecturer – 'Always stick to what the system provides, otherwise you may get into trouble'. It mystifies the computer system by putting its most vital functions into a software package which is beyond the control and comprehension of the applications programmer, thus introducing even into the exclusive province of Data Processing the division between software experts and other programmers, and reinforcing the idea that we do not really control the tools we use, but can only do something 'if the operating system lets you' – a phrase which I am sure many of us have used. The system which results seems absurdly top heavy and complex. The need to have everything centrally controlled seems to impose an enormous strain."

The introduction of computerised systems is frequently used as a smokescreen to introduce another management control weapon – job evaluation. Pseudo-scientific reasons are given for fragmenting jobs and slotting the subdivided function into a low level of the system hierarchy with correspondingly low wages for 'appropriate' job grades. My experience of this in industry tends to show that it is frequently actually used to consolidate the unequal pay and opportunities of women by either implying (they can no longer state it openly as they have done in the past), or ensuring by structural means and recruitment, that the fragmented functions are 'women's work'. There is of course no such thing as 'women's work' in this sense any more than there is women's mathematics, women's physics, women's literature or women's music. There is only work – the means by which employers extract profits from all of us but higher profits from women. Thus

Contradictions of Science and Technology

a contradiction will exist in that although scientific and technological progress could provide the objective circumstances for greater equality between the sexes in the productive process, in our profit-orientated society the reverse will frequently be the case. Women will have to fight not only the traditional forms of discrimination, but much more sophisticated and scientifically structured ones with little indication that the unions catering for such workers have really understood the nature or scale of these problems.

The emergence of fixed capital as a dominant feature in the productive process means that the organic composition of capital is increased and industry becomes capital intensive rather than labour intensive. Human beings are increasingly replaced by machines. This in itself increases the instability of capitalism; on the one hand, capitalism uses the quantity of working time as the sole determining element, yet at the same time continuously reduces the amount of direct labour involved in the production of commodities. At an industrial level, literally millions of workers lose their jobs and millions more suffer the nagging insecurity of the threat of redundancy. An important new political element in this is the class composition of those being made redundant. Just as the use of high-capital equipment has spread out into white-collar and professional fields so also the consequences of high-capital equipment do likewise. Scientists, technologists, professional workers, clerical workers, all now experience unemployment in a manner that only manual workers did in the past. Verbal niceties are used to disguise their common plight. A large West London engineering organisation declared its scientists and technologists 'technologically displaced', its clerical and administrative workers 'surplus to requirements', and its manual workers 'redundant'. In fact they had all got the good old-fashioned sack! In spite of different social, cultural and educational backgrounds, they all had a common interest in fighting the closure of that plant, and they did. Scientists and technologists paraded around the factories carrying banners demanding 'the right to work' in a struggle that would have been inconceivable a mere ten

The Search for Alternatives

years ago. Technological change was indeed proletarianising them. In consequence of the massive and synchronised scale of production which modern technology requires, redundancies can affect whole communities. During a recession in the American aircraft industry in the early 1970s a union banner read: 'Will the last person leaving Seattle please turn out the lights.' Because of this change in the organic composition of capital, society is gradually being conditioned to accept the idea of a permanent pool of unemployed people. Thus we find in the United States, in spite of the artificial stimulus of the Vietnam War, over the past ten years about five million people have been permanently out of work.

We have witnessed in this country the large-scale unemployment of recent years. Unemployment is considerable in Italy, and even in the West German miracle there are sections of workers – particularly over the age of fifty – who are now experiencing long terms of unemployment. This unemployment itself creates contradictions for the ruling class. It does so because people have a dual role in society, that of producers and consumers. When you deny them the right to produce, you also limit their consumption power. In an attempt to achieve a balance, efforts are now being made to restructure the social services to maintain that balance between unemployment and the purchasing power of the community. In the United States, President Kennedy spoke of a 'tolerable level of unemployment'. In Britain in the 1960s Harold Wilson, stoking the fires of industry with the taxpayer's money through the Industrial Reorganisation Corporation to create the 'white heat of technological change', spoke in a typical double negative of a 'not unacceptable level of unemployment'.

A remarkable statement for a so-called socialist Prime Minister! The net result is that there is on the one hand an increased work tempo for those in industry, whilst on the other hand there is a growing dole queue, with all the degradation that that implies; nor is there any indication that the actual working week has been reduced during this period. Indeed, in spite of all the technological

Contradictions of Science and Technology

change since the war, the actual working week in Britain for those who have jobs is longer now than it was in 1946. Yet the relentless drive goes on to design machines and equipment which will replace workers. Those involved in such work seldom question the nature of the process in which they are engaged. Why, for example, the frantic efforts to design robots with pattern recognition intelligence when we have a million and a half people in the dole queue in Britain whose pattern recognition intelligence is vastly greater than anything yet conceived even at a theoretical level?

The system seeks in every way to break down the workers' resistance to being sacked. One of the sophisticated devices was the Redundancy Payments Act under the 1964-70 Labour Government. Practical experience of trade unions in Britain demonstrates that the lump sums involved broke up the solidarity at a number of plants where struggle was taking place against a closure.

A much more insidious device is to condition workers into believing that it is their own fault that they are out of work, and that they are in fact unemployable. This technique is already widespread in the United States, where it is asserted that certain workers do not have the intelligence and the training to be employed in modern technological society. This argument is particularly used against black workers, Puerto Ricans and poor whites. There is perhaps here fertile ground for some of the 'objective research' of Jensen and Eysenck.

The concept of a permanent pool of unemployed persons as a result of technological change also brings with it the danger that those unemployed would be used as a disciplining force against those still in work. It undoubtedly provides a useful pool from which the army and police force can draw, and during the recent redundancies in Britain, a considerable number of redundant workers from the North-east were recruited into the army and then used against workers in Northern Ireland. Coupled with the introduction of this high-capital equipment is usually a restructuring known as 'rationalisation'. The epitome of this in

The Search for Alternatives

Britain is the GEC complex with Arnold Weinstock at its head. In 1968, this organisation employed 260,000 workers and made a profit of £75 million. In consequence of quite brutal redundancies, the company's work force was reduced to 200,000, yet profits went up to £105 million. These are the kind of people who are introducing high capital equipment, and they make their attitude to human beings absolutely clear. It is certainly profits first and people last! One quotes Arnold Weinstock not because he is particularly hideous (he is in fact extremely honest, direct and frank) but because he is prepared to say what others think. He said on one occasion that 'People are like elastic, the more work you give them, the more they stretch.' We know, however, that when people are stretched beyond a limit, they break. My union has identified a department in a West London engineering company where the design staff were reduced from thirty-five to seventeen and there were six nervous breakdowns in eighteen months. Yet people like Weinstock are held up as a glowing example to all aspiring managers. One of his own senior managers once boasted that 'he takes people and squeezes them till the pips squeak'. I think it is a pretty sick and decaying society that will boast of this kind of behaviour.

Most industrial processes, however capital intensive they might be, still require human beings in the total system. Since highly mechanised or automated plant frequently is capable of operating at very high speeds, employers view the comparative slowness of human beings in their interaction with the machinery as a bottleneck in the overall system. In consequence of this, pay structures and productivity deals are arranged to ensure that the workers operate at an ever faster tempo.

In many instances the work tempo is literally frantic. In one automobile factory in the Midlands in Britain they reckon that they 'burn a man up' on the main production line in ten years. They recently tried to get our union to agree that nobody would be recruited for this type of work over the age of thirty. For the employer it is like having a horse or dog. If you must have one at all, then you have a young one so that it is energetic and frisky

Contradictions of Science and Technology

enough to do your bidding all the time. So totally does the employer seek to subordinate the worker to production, that he asserts that the worker's every minute and every movement 'belong' to the employer. Indeed, so insatiable is the thirst of capital for surplus value, that it thinks no longer in terms of minutes of workers' time, but fractions of minutes. The grotesque precision with which this is done to workers can be seen from a report which appeared in the *Daily Mirror* of 7 June 1973. It gave particulars of the elements which make up the 32.4 minute rest-allowance deal for body press workers on the Allegro car: trips to the lavatory 1.62 minutes (note; not 1.6, not 1.7, but 1.62!); recovery from fatigue 1.30 minutes, sitting down after standing too long sixty-five seconds, for monotony thirty-two seconds. The report went on to point out that, in a recent dispute, the workers sought to increase the monotony allowance by another sixty-five seconds! The methods may vary from company to company, or from country to country, but where the profit motive reigns supreme, the degradation and subordination of the worker is the same. George Friedmann has written of two different methods used by great French companies, Berliot in Lyons and Citroen in Paris:

> "Why has the Berliot works the reputation, in spite of the spacious beauty of its halls, of being a jail? Because here they apply a simplified version of the Taylor method of rationalising labour, in which the time taken by a demonstrator, an 'ace' worker, serves as the criterion imposed on the mass of workers. He it is who fixes, watch in hand, the 'normal' production expected from a worker. He seems when he is with each worker, to be adding up in an honest way the time needed for the processing of each item. In fact if the worker's movement seem to him to be not quick or precise enough, he gives a practical demonstration, and his performance determines the norm expected in return for the basic wage. Add to this supervision in the technical sphere the disciplinary supervision of uniformed warders to patrol the factory all the time and go as far as to push open doors of the toilets to check that the men squatting there are not smoking, even in workshops where the risk of fire is non-existent. At Citroen's the methods used are

The Search for Alternatives

more subtle. The working teams are in rivalry with one another, the lads quarrel over travelling cranes, drills, pneumatic grinders, small tools. But the supervisors in white coats, whose task is to keep up the pace, are insistent, pressing, hearty. You would think that by saving time a worker was doing them a personal favour. But they are there, unremittingly on the back of the foreman, who in turn is on your back; they expect you to show an unheard of quickness in your movements, as in a speeded-up motion picture! Within this context, the desire of companies to recruit only those under the age of 30 can be seen in its dehumanised context."[3]

Although this is the position on the workshop floor, it would be naive indeed to believe that the use of high-capital equipment will be any more liberating in the fields of clerical, administrative, technical, scientific and intellectual work.

Age limits are now gradually being introduced in the white-collar areas. In 1971 the *Sunday Times* gave a list of the peak-performance ages for mathematicians, engineers, physicists and others. For some of these the peak-performance age was twenty-nine and thirty. It has been suggested that in order to utilise this high-capital equipment as effectively as possible, a careers profile should be worked out for those who have to interface with it.

When workers reach their peak-performance age, it is suggested that this should be followed by a careers plateau for three or four years and thereafter, unless the employee has moved into management, that they be subjected to a 'careers de-escalation'. The obvious extension of the careers de-escalation is redundancy. Practical experience demonstrates, particularly during periods of redundancy, that older people are being eliminated in this way. They are being eliminated or down-graded to lower-paid work simply because they have committed the hideous crime of beginning to grow old. We are, as Samuel Beckett once said, 'all born of the gravedigger's forceps'. Growing old is the most natural human process. It is a biological process, but, in the contradictory nature of our profit-orientated society, it is treated almost as a crime. It is true that the kind of equipment we have been discussing imposes very stringent demands upon those who have to interface

Contradictions of Science and Technology

with it. Seen in terms of the total man/machine systems, people are slow, inconsistent, unreliable, but still highly creative. The machine is the dialectical opposite, in that it is fast, reliable, consistent, but totally non-creative. As people attempt to respond to the machine, enormous stress is placed upon them. My union has identified areas within the design activity where by using interactive graphic systems the decision-making rate of the designer is increased by 1900 per cent.

HUMANS AS MACHINE

Again there are analogies to be drawn from the shop floor. In the British Steel Corporation a productivity agreement has introduced medical checks. In practice these medical checks meant the operators were tested to ensure that their response rates were fast enough to interface with the equipment. They were merely tested for their response rates as a diode might be. A series of occupational suitability tests and character compatibility assessments are now gradually being used to do the same sort of thing to white-collar workers who have to use high-capital equipment. The object is to transform the worker, whether by hand or brain, into a suitable machine appendage. To do this, all the human requirements of the individual must be denied. They must be transformed into operating units. The 'scientific' manner in which this man/ machine interface is planned emphasises the total dehumanisation of the so-called technologically advanced production techniques. Robert Boguslaw has recently pointed out how some behavioural scientists view the human being in this situation:

> "Our immediate concern let us remember, is the explication of the operating unit approach to system design, no matter *what* materials are used. We must take care to prevent this discussion from degenerating into a single-sided analysis of the complex characteristics of one type of system material: namely human beings. What we need is an inventory of the way in which human behaviour can be controlled and

The Search for Alternatives

a description of some instruments which will help us to achieve control. If this provides us with sufficient 'handles' on human materials so that we can think of them as one thinks of metal parts, electrical power, or chemical reactions, then we have succeeded in placing human material on the same footing as any other materials and can proceed with our problems of systems design."[4]

This, then, is the objective dehumanisation which takes place side by side with this advanced technology.

"There are however, many disadvantages in the use of these human operating units. They are somewhat fragile; they are subject to fatigue, obsolescence, disease and death; they are frequently stupid, unreliable and limited in memory capacity. But beyond all this, they sometimes seek to design their own circuitry. This, in a material, is unforgiveable. Any system utilising them must devise appropriate safeguards."

Thus, if workers use their greatest attribute – that is, their ability to think – their ability to design their own circuitry – this is regarded as disruptive. The objective requirement of industry, then, is for people who will act as robots, people who are interchangeable with robots. Some scientists and technologists take the smug view that this can only happen in any case to mere manual workers on the shop floor. They fail to realise that the problem is now at their own doorstep. At a conference on robot technology at Nottingham University in April 1973, a programmable draughting or design system was accepted by definition as being a robot. One of the manufacturers of robotic equipment pointed out that 'Robots represent industry's logical search for an obedient workforce.' This is a very dangerous philosophy indeed. The great thing about people is that they are sometimes disobedient. Most human development, technical, cultural and political, depended upon those movements which questioned, challenged and where necessary disobeyed the then established order.

The ruling class views all workers, whether by hand or brain, as units of production. Only when that class reality has been firmly grasped can the chasm which divides the potentialities of science and technology from the current reality be understood. The gap between that which is possible, and that which is, widens daily. Technology can produce a Concorde but not enough simple heaters to save the hundreds of old-age pensioners who each winter die in London of hypothermia. Only when one realises that the system regards old-age pensioners as discarded units of production does this make sense – capitalist sense. This is part of their social design, and from a ruling-class viewpoint is quite 'scientific' and abides closely by the principles observed in machine design. I know, as a designer, that when you design a unit of production you ensure that you design it to operate in the minimum environment necessary for it to do its job. You seek to ensure that it does not require any special temperature-controlled room unless it is absolutely essential. In designing the lubrication system you do not specify any exotic oils as lubricants unless it is necessary. You ensure that its control system is provided with the minimum brain necessary for it to do its job. You do not, for example, have a machine tape-controlled if you can get away with a manual one. Finally, you provide it with the minimum amount of maintenance; in other words, you design for it the maximum life span in which it will operate before a failure.

Those who control our society see human beings in the same way. The minimum environment for people means that you provide them with the absolutely lowest level of housing which will keep them in a healthy enough state to do their job. If one doubts that, it is still worth remembering that 7,000,000 people live in slums in Britain. The equivalent of fuel and lubrication for the machine is the food provided for a person. This is also kept at a minimum for those who work. We even find Oxford dieticians still telling old-age pensioners how they can manage on £2 of food per week. The minimum brain is provided by an educational system which gives people enough knowledge to do their job, which trains them to do their job, but does not educate them to think about their predicament or that of society as a whole.

The minimum maintenance necessary is provided through the

The Search for Alternatives

National Health Service, which concentrates on curative rather than preventive medicine, and the reality, the harsh reality, is that when people have finished their working life, they are thrown on the scrap heap like an obsolete machine.

If that is felt to be an extreme position, it is worth recalling the statement of the doctor at Willesden Hospital, who said there was no need for National Health patients over the age of sixty-five to be resuscitated (the doctor was actually sixty-eight!). When a barrage of protest was raised the statement was hurriedly withdrawn as a mistake! The real mistake he made was to reveal in naked print one of the underlying assumptions of our class-divided society. Science and technology eannot be humanely applied in an inherently inhuman society, and the contradictions for scientific workers in the application of their abilities will grow and, if properly articulated, will lead to a radicalisation of the scientific community.

A source of great stress, particularly for white-collar workers, is the problem of knowledge obsolescence. This problem is closely related to the rate at which technology itself is changing. It seems desirable to attempt to quantify technological change. The scale of technological development in the last twenty years is probably equal to that in all of humanity's previous existence.

The scale of scientific effort, which is closely related to technological change, has in the present century increased out of all recognition. Bernal calculated that in 1896 there were perhaps in the whole world some 50,000 people who between them carried on the whole tradition of science, not more than 15,000 of whom were responsible for the advancement of knowledge through research. Today, the total number of scientific workers in industry, government and academic circles in Britain alone is over 400,000. This is merely a reflection in manpower of the statistics of the actual rate of technological change, which in the last century alone has meant that our speed of communication was increased by 10^7, our speed of travel by 10^2, data-handling by 10^6, energy resources by 10^3 and weapons power by 10^6.

As the rate of technological change increases, so also does the

Contradictions of Science and Technology

rate at which knowledge becomes obsolete. Mathematical models described by Sir Frederick Warner indicate that in order to keep abreast of this knowledge engineers would have to spend 15 per cent of their time in up-dating their current knowledge. Mr Norman McRae, Deputy Editor of *The Economist,* stated in the January 1972 issue that

> "The speed of technological advance has been so tremendous during the past decade that the useful life of knowledge of many of those trained to use computers has been about three years. [and, further] A man who is successful enough to reach a fairly busy job at the age of 30, so busy that he cannot take sabbatical periods for study, is likely at the age of 60 to have about one-eighth of the scientific (including business scientific) knowledge that he ought to have for the proper functioning in his job."

It has even been suggested that if one divided knowledge into quartiles of out-datedness, those in the age bracket over forty-five would find themselves in the same quartile as Pythagoras and Archimedes. The stress that this places upon staff workers, in particular older people, should not be underestimated. It is the responsiblity of the trade unions to protect these older people. This they should do not in any patronising, benevolent fashion, but in recognition of the class right of these older people to work at a civilised tempo. For these are the ones who in the past have created the real wealth that has made the purchase of this kind of high-capital equipment possible. All younger technologists should fully understand that however energetic and forceful they may feel now, they will inevitably begin to grow old, and if they allow older members to be treated in this way they are creating a framework of oppression which will be used against them in the future.

THE FRAGMENTATION OF SKILLS

A major part of the process of technological change is the fragmentation of jobs into deskilled, narrow elements. It is also

The Search for Alternatives

part of the historical division between intellectual and manual work. In the past, many jobs which were essentially manual did contain within them major elements of intellectual and scientific work. Sir William Fairbairn's definition of a millwright in 1861 illustrates the point.

> "The millwright of former days was to a great extent the sole representative of mechanical art ... he was an itinerant engineer and mechanic of high reputation. He could handle the axe, the hammer and the plane with equal skill and precision; he could turn, bore or forge with the despatch of one brought up to these trades and he could set out and cut furrows of a millstone with an accuracy equal or superior to that of the miller himself .... Generally he was a fair mathematician, knew something of geometry, levelling and mensuration, and in some cases possessed a very competent knowledge of practical mathematics. He could calculate the velocities, strength and power of machines, could draw in plan and section, and could construct buildings, conduits, or water course in all forms and under all conditions required in his professional practice. He could build bridges, cut canals and perform a variety of work now done by civil engineers."[5]

All the intellectual work has been long since withdrawn from the millwright's function.

This fragmentation of skills now applies equally in the white-collar areas. The draughtsman of the 1930s in Britain was the centre of design. He could design the component, stress it, specify the materials to be used, define the method of lubrication, and write the test specifications. With the increasing complexity of technology, each of these have now been fragmented into narrow, specialised areas. The draughtsman draws, the stressmen carry out the calculations, the metallurgist specifies the materials, the tribologist decides upon the means of lubrication.

It has been common for some time to talk about 'dedicated machines'. It is now a fact that when defining a job function employers define a dedicated appendage to the machine, the operator.

Even our educational system is being distorted to produce these

Contradictions of Science and Technology

'dedicated people for dedicated machines'. It is no longer a matter that the people are being educated to think; they are being trained to do a narrow, specific job. Much of the unrest amongst students is recognition that they are being trained as industrial fodder for the large monopolies in order to fit them into narrow fragmented functions where they will be unable to see in an overall panoramic fashion the work on which they are engaged.

In order to ensure that the right kind of 'dedicated product' is turned out of the university, we find the monopolies attempting to determine the nature of university curricula and research programmes. Warwick University was a classical example. In particular, at research level, the monopolies increasingly attempt to determine the nature of research through grants which they provide to universities or research projects undertaken in their own laboratories. Many research scientists still harbour illusions that they are in practice 'independent, dedicated searchers of truth'.

The 'truth' for them has to coincide with the interests of the monopolies if they are to retain their jobs. William H. Whyte Jr pointed out that in the United States, out of 600,000 persons then engaged in scientific research, not more than 5000 were allowed to choose their research subject and less than 4 per cent of the total expenditure was devoted to 'creative research', which does not offer immediate prospects of profits.

He recognises the long-term consequences of this and concludes:

> 'If corporations continue to mould scientists the way they are now doing, it is entirely possible that in the long run this huge apparatus may actually slow down the rate of basic discovery it feeds on.'[6]

PERSPECTIVES FOR REVOLUTIONARY ACTION

I have up to now concentrated on the contradictions as they affect the worker by hand or brain. There are of course problems for the employer, and an understanding of some of these is of considerable tactical importance.

The Search for Alternatives

One of the contradictions for the employer is that the more capital he accumulates in any one place, the more vulnerable it becomes. The more closely he synchronises his industry and production by using computers, the greater becomes the strike power of those employed in it. Mao Tse Tung once said, in his military writings, that the more capitalised an army becomes, the more vulnerable it becomes also. This has been demonstrated in Vietnam, where NLF cadres with £1.50 shells were able to destroy American aircraft with airborne computers costing something like £2.5 million each. A Palestinian guerilla with a revolver costing perhaps £20 can hijack a plane costing several million dollars and destroy it at some safe airfield. High-capital equipment, although it appears all powerful and invincible, always has a point of vulnerability, and possibilities for sabotage and guerilla warfare are considerable. A quite small force can destroy or immobilise plant equipment or weapons costing literally millions. The capitalisation of industry also produces an analogous situation. In the past, when a clerical worker went on strike, it had precious little effect. Now if the wages of a factory are carried out by a computer, a strike by clerical workers can disrupt the whole of the plant. It is also true on the factory floor that in the highly synchronised motor-car industry, a strike of twelve workers in the foundry can stop large sections of the entire motor-car industry.

The same is happening in the design area. As high-capital equipment, through Computer Aided Design, is being made available to design staffs, first it proletarianises them, but second it also increases their strike power. In the past, when draughtsmen went on strike, they simply put down their 6H pencils and their rubbers, and there was unfortunately a considerable length of time before an effect was felt upon production, even when the manual workers were blacking their drawings. With the new kind of equipment described, where computer tapes are being prepared or where high-capital equipment is used for interactive work, the effects of a strike will in many instances be immediate, and production will be affected in a very short length of time.

This will apply equally to hosts of other jobs and occupations,

Contradictions of Science and Technology

in banking, insurance, power generation, civil transport, as well as those more closely connected with industry and production.

Thus, whilst the introduction of fixed capital enables the employer to displace some workers and subordinate others to the machine, it also embodies within it an opposite in that it provides the worker with a powerful industrial weapon to use against the employer who introduced it.

This is even the case when industrial action short of strike action is taken. As has been pointed out, the activity of the worker is transformed to suit the requirement of fixed capital. The more complete that transformation, the greater is the disruptive effect of the slightest deviation by the worker from a predetermined work sequence. Industrial militants with an imaginative and creative view of industrial harassment have been able to exploit this contradiction by developing such techniques: 'working to rule', 'working without enthusiasm' and 'days of non-co-operation'. I know from personal experience that these techniques can reduce the output of both manual and staff workers by up to 70 per cent without placing on the workers involved the economic hardship of a full strike.

Since much of the sophisticated equipment I have described earlier is very sensitive and delicate in a scientific sense, it has to be handled with great care and is accommodated in purpose-built structures in conditions of clinical cleanliness. In many industries the care the employer will lavish on 'his' fixed capital is in glaring contrast with the comparatively primitive conditions of 'his' living capital. The campaign for parity with equipment, which perhaps started facetiously in 1964 with that placard at Berkeley which parodied the IBM punchcard ('I am a human being: Please do not fold, spindle or mutilate') has now assumed significant industrial dimensions. In June 1973 designers and draughtsmen members of the AUEW-TASS employed by a large Birmingham engineering firm, officially claimed 'Parity of environment with the CAD Equipment' in the following terms:

"This claim is made in furtherance of a long standing complaint

The Search for Alternatives

concerning the heating and ventilation in the Design and Drawing Office Area going back to April 1972. Indeed to our certain knowledge these working conditions have been unsatisfactory as far back as 1958. We believe that if electromechanical equipment can be considered to the point of giving it an air conditioned environment for its efficient working the human beings who may be interfaced with this equipment should receive the same consideration."

It is an interesting reflection on the values of advanced technological society that it subsequently took three industrial stoppages to achieve for the designers, conditions approaching those of the CAD equipment. The exercise also helped to dispel some illusions about highly qualified design staff not needing trade unions.

Scientists must now begin to learn the lessons of such experiences, and to understand that their destiny is bound up with all of those 'moulded' by the systems. Only when they are prepared to be involved in political struggle with them, can they ever begin to move towards a society where scientists will be able to give 'according to their ability'. It is the historical task of the working class to effect such a transformation, but in that process scientists and technologists can be powerful and vital allies for the working class as a whole. This will mean that scientists will have to involve themselves in the political movement. Above all, they must attempt to understand that the products of their ingenuity and scientific ability will become the objects of their own oppression and that of the mass of the people until they are courageous enough to help form that sort of society. Karl Marx, writing in his *Critique of the Gotha Program*, argues:

> "When the enslaving subordination of the individual to the division of labour, and with it the antithesis between mental and physical work has vanished; when labour is no longer merely a means of life but has become life's principal need; when the productive forces have also increased with the all-round development of the individual and all the springs of co-operative wealth flow more abundantly. Only then will it be possible completely to transcend the narrow outlook of bourgeois

Contradictions of Science and Technology

right and only then will society be able to inscribe on its banners: From each according to his ability, to each according to his needs."[7]

Then, and then only will scientists be able to truly give of their ability to meet the needs of the community as a whole rather than maximise profits for the few.

Notes

1. K. Marx, *Capital,* vol. I (London: Lawrence & Wishart, 1974) p. 174.
2. Quoted in *Realtime,* 6 (1973).
3. G. Friedmann; quoted in E. Mandel, *Marxist Economic Theory* (London: Merlin Press, 1971) p. 183.
4. R. Boguslaw, *The New Utopians: a Study of System Design and Social Change* (Englewood Cliffs, N.J.: Prentice-Hall, 1965).
5. W. Fairbairn; quoted by]. B. Jefferys, *The Story of the Engineers* (London: Lawrence & Wishart for the AEU, 1945) p. 9.
6. W. H. Whyte, *The Organisation Man* (New York: Simon & Schuster, 1956).
7. K. Marx, *Critique of the Gotha Programme,* ed. C. P. Dutt (London: Lawrence & Wishart, 1938).

First published:

The Political Economy of Science
Steven and Hilary Rose (Editors)
MacMillan, 1976

THE SEARCH FOR ALTERNATIVES

First published in *Defence Cuts and Labour's Industrial Strategy*
Labour CND Pamphlet, 1976

For as long as I can recall, defence cuts have been part of the policies of the Labour left and large sections of the trade union movement. The argument in support of this has been that defence cuts would release capital, which could then be used in the social services. It is of course then grudgingly admitted that there would be the residual problem of unemployment.

That this should be so reveals the extent to which we have been conditioned by the criteria of the market economy. Thus we see the freeing of capital as an asset, and the freeing of people as a liability. In doing this we ignore our most precious asset – people, with their skill, ingenuity and creativity.

In the defence and aerospace industries, we have some of the most highly skilled and talented workers in this country. Yet, like the ruling class, we have thought of capital first and people second, and ignored the incredible contribution which these people's skill and ability could make to the economy as a whole and the wellbeing of the people of this country.

Capital is merely paper money. You can't eat a pound note, you can't drive round in one, you can't live in one. It only has its meaning if people like you and I go into factories day in, day out and create the real wealth that money represents. It is this real wealth which is important to the nation.

It seems to me to be a measure of the bankruptcy of our political and trade union leaders that when they talk about defence cuts,

The Search for Alternatives

they have no concrete proposals whatsoever about alternative work. There are of course vague generalisations about those who are displaced finding alternative work in the social services. There is no planned transition from one to the other, and in fact because of Government policies there are large pools of unemployment in the service industries in any case. Such jobs as are vacant certainly will not attract skilled aerospace workers. If anybody thinks that a skilled prototype fitter from my factory is going to take a job as a hospital porter, they seriously misunderstand workers' aspirations.

THE LUCAS CORPORATE PLAN

In the absence of any concrete proposals for alternative work, defence and aerospace workers have repeatedly found themselves in the position of either demanding that military projects be continued or facing the dole queue. I do not accept that this is the choice, any more than I accept that the choice recently facing the Chrysler workers was between the dole queue and producing rubbishy Chrysler cars. We in Lucas Aerospace have attempted to demonstrate, with our corporate plan, that there are other real alternatives if we seriously address ourselves to the problem.

At present Lucas Aerospace is heavily dependent on military work: 43% of its business is military aircraft projects, and a further 7% is other defence work. It will be heavily involved in the new Tornado multi-role combat aircraft, which is likely to cost Britain at least £4,000 million.

The corporate plan, detailing alternatives to this dependence on military work, was produced by the Lucas Aerospace Combine Shop Stewards Committee. This body is unique in the British trade union movement, in that it represents all 14,000 workers in the 17 British sites of Lucas Aerospace. More importantly, it organises everybody from the highest level technologist to the semi-skilled worker. Thus it brings together on the one hand the analytical power of the technologist but on the other, and perhaps more importantly, the class understanding of those on the shop floor.

It is only possible to generate a corporate plan and have the

The Search for Alternatives

power to fight for it when one has established a strong and powerful organisation at the point of production, as we have in Lucas Aerospace. The Combine Committee grew up in the late 1960s to resist the sackings which were the direct consequence of Wilson's white heat of technological change, which we found simply to be burning up jobs. The Combine Committee is now well enough organised to prevent the company engaging in any direct sackings, but we have not succeeded in preventing the company running down sectors of the workforce by so-called natural wastage.

PRODUCT PROPOSALS

1. Oceanic equipment – for use in the exploration and extraction of natural gas, collection of mineral-bearing nodules from the sea bed, and submarine agriculture.

2. Telecheiric machines – electro-mechanical extensions to the human body, remotely controlled by the operator, for use in dangerous environments.

3. Transport systems – lightweight road/rail vehicles; hybrid internal combustion/ battery-powered vehicles, combining the best characteristics of both; airships.

4. Braking systems – safe systems for both road and rail vehicles.

5. Alternative energy sources – wind generators; solar collectors, producing electrical output or direct heating; tidally-driven turbines.

6. Medical equipment – portable life support systems for ambulances; kidney machines; aids for the disabled; sight-substituting aids for the blind.

7. Auxiliary power units – interchangeably driven by petrol, diesel or methane, and able to operate as a pump, compressor or generator.

8. Micro-processors – electronic devices for continuously monitoring and controlling the operation of large machines.

9. Ballscrews – used for converting rotating to linear motion, or *vice versa,* with wide applications to machine tools and other products in the plan.

The Search for Alternatives

There are of course many contradictions which demonstrate the absurdities of our so-called technologically advanced society. Two of them were particularly important in stimulating our members to consider the corporate plan. First, there is the enormous gap between that which technology could provide and that which it does provide. We have a level of technology which can produce Concorde, yet at the same time old age pensioners are dying of hypothermia through lack of simple urban heating systems – I understand that something like 980 died in the winter of 1975/6 in the London area alone. In the automotive industry we have senior design engineers who optimise the body configuration of cars such that they are aerodynamically stable at 100 miles per hour. Yet the average velocity of cars through New York is 6.2 mph. In fact at the turn of the century, when they were horse-drawn, it was something like 11 mph. The same kind of absurd situation is beginning to develop in cities in this country.

Secondly, we have a dole queue of about 1.25 million, and if one took into account women workers and those who wish to work part time, this figure is really more like 1.8 million. Amongst these are thousands of highly skilled engineers when we urgently need cheap urban transport systems. Thousands of electricians are in the dole queue, when we need cheap, effective heating systems. And there are thousands of building workers in the dole queue, when seven million people live in semi-slums in this country, and there is a tragic need for more hospitals and schools.

Taking both these contradictions together, the workforce in Lucas Aerospace has put forward in its corporate plan a demand for the right to use our skill and ability on socially useful products. We are demanding not merely the right to work, but the right to work on products which will be in the interest of the community as a whole.

In drawing up our corporate plan, we sent out 180 letters to institutions, parties, trade union and individuals, asking them if they had suggestions of products on which we might work. It was a bitter disappointment to us, but an important political experience, to learn that those who had been making profound speeches in the

The Search for Alternatives

cosy sanctuary of seaside conference halls were quite unable to make any concrete suggestions. In fact only three positive proposals were submitted to us, from Dr Elliott at the Open University, Professor Thring at Imperial College, and Richard Fletcher and Clive Latimer at North East London Polytechnic.

In the absence of any real proposals, we then sent out a questionnaire to all our shop stewards committees throughout the country. The questionnaire was deliberately designed in such a fashion as to cause our members to become more conscious of their skills, and of the facilities and machine tools in their plants – facilities, incidentally, which had been paid for as a result of the profits – which they had helped to make for the company in the past.

As a result of the questionnaire, we received some 150 proposals for alternative products on which we might work. We selected from these 12 product areas which related directly to the skill and ability of our members and to the manufacturing facilities at each site. We also selected them to achieve a balance of long and short term projects, projects of direct use in this country and for newly emergent nations, and projects which required a high capital investment and those which could be produced almost straight away.

A full summary of the products included in the corporate plan is given in the panel, and here I will simply comment on some of them.

We are firstly considering components for *low energy housing.* Amongst these are solar heating systems, in particular the switching circuits, and a range of heat pumps which could be used in conjunction with them. We would like to see our experience in the field of aerodynamics used to produce large scale wind generators. In some of these systems, instead of converting the energy inefficiently by way of electricity, we have ideas about using the wind generator to drive a series of pumps which would heat the liquid directly. We could also arrange for rotor speeds to vary according to wind velocity, giving a constant power output. We are particularly keen to see that any developments in this area

The Search for Alternatives

would be used for community heating systems rather than as gimmicks for individual architect-built houses, since there is a tendency for certain sections of alternative technology to be used as playthings for the middle classes.

Over the years we have developed considerable expertise in the field of dynamometry. We feel that this talent should be used to produce auxiliary ('fail-safe') *braking systems* for cars, coaches and trains. In fact it has already been suggested that if a device of this kind had been fitted to British coaches, the coach disaster in Yorkshire in 1975 would not have taken place.

We are particularly interested to examine alternative *transport systems*. One of our proposals in this field is for a *hybrid road/rail vehicle*. Basically this would be a vehicle with a very light structure which would run on pneumatic-tyred wheels. It would be capable of travelling through cities as a coach, and then driving onto railway track, making full use of the magnificent railway network we have in this country.

But not merely would this vehicle be of use in Britain; it could make an enormous contribution in developing countries. With a traditional piece of railway rolling stock, you have a metal rim running on a metal track. Because of the low friction involved this is not capable of going up an incline steeper than 1 in 80. This means that when new railways are laid down, it is necessary literally to flatten the mountains and fill the valleys, so that you end up with an extremely level track. Our vehicle would be capable of climbing gradients as steep as 1 in 6 or 1 in 8. Thus instead of being confronted with about £1 million per track mile, which was the cost of the new railway line in Tanzania, with our vehicle a very light track could simply follow the contours of the existing terrain, and would cost only £20,000 per track mile. The Highlands and Islands Development Board is already showing a keen interest in this vehicle for the enormous role it could play in developing a socially desirable transport system in Scotland. Since the late 1940s the Labour Party has had a policy for an integrated transport system; this vehicle could play an enormous part in such a policy.

The Search for Alternatives

Also in the field of transport systems, we are proposing a quite revolutionary *hybrid power pack*. Basically this would make the best use of the characteristics of both the internal combustion engine and the electric motor. Much has been said recently about battery-powered vehicles. It is true that ecologically they are highly desirable. On the other hand, until there is an enormous breakthrough in battery technology, and I don't see this in the immediate future, it will be necessary to stop and charge a vehicle on a stop-start journey every 40 miles, and on a continuous journey over flat terrain every 100 miles.

With the power pack we are talking about, there is an internal combustion engine running at its constant optimum speed, driving a generator which charges a stack of batteries, and that in turn drives the electric motor. Because the greatest wastage of fuel is in stopping and starting, in accelerating and decelerating, the fuel consumption of this vehicle would be about 50% less than a normal car. Also, the toxic emissions which tend to occur during idling, stopping and starting, changing gear, etc, would be reduced by some 80%. An additional advantage of this vehicle is that it would be very quiet: a similar power pack which we've tested is almost inaudible against normal background noise at a distance of 10 metres. Thus one would have a power pack which would be very quiet, would save fuel, and would enormously reduce toxic emissions. But clearly, a unit of this kind would mean a radical re-examination of the automotive industry, and the throw-away philosophy which underlies it.

Another proposal is for a range of *auxiliary power units* for developing countries.

When the gin-and-tonic brigade go to these countries to sell them power packs, they try to sell them one which is suitable for only one application. In our case we're suggesting a basic prime mover, which could be adapted for the type of fuel available in a particular area, and which instead of being simply an electrical generator would have a series of alternative heads, which could provide interchangeably electrical output, hydraulic output or compressed air, or could even be used as a pumping arrangement.

The Search for Alternatives

Thus a village in a remote area could have a single unit which could meet a number of its requirements, rather than purchasing a separate unit for each of these.

We are proposing a number of products in the field of *medical engineering*. We already make pacemakers and kidney machines. In fact we were recently engaged, through the unions and the local Trades Council, in preventing the company selling off its kidney machine division to a large international monopoly. We regard it as scandalous that people are dying for want of a kidney machine, when those who could be making them are actually facing the prospect of redundancy. Certainly those displaced in the aerospace industry could in large numbers be engaged in this type of work.

We feel that considerable use will be made in coming years of *oceanic equipment,* to exploit the resources of the sea bed. We must say we view this with some trepidation, bearing in mind the wanton and depraved way humanity has used the surface of the earth. However, in this field we are talking about a range of submersible vehicles which could be used for marine agriculture and the collection of metal-bearing nodules, without human beings having to submit themselves to the hazards of the depths involved.

LABOUR-INTENSIVE WORK

Directly related to this product range is what we call *telecheiric devices.* These are pieces of equipment which mimic, in a remote environment, the actual motions and actions of a human being. We are particularly keen on the philosophy underlying these devices since, unlike robotic equipment, which takes part of the skill of a human being and objectifies it in some kind of memory or magnetic tape (and in consequence frequently leads to redundancies), in these telecheiric devices the human beings remain in full control all the time. Thus this is a way of blending high technology with a relatively labour-intensive form of work.

It has recently been suggested that robotic devices should be used to maintain North Sea oil equipment. In this way the incredible hazards involved might be reduced. It has to be

The Search for Alternatives

recognised, however, that to programme a robot to perform even very simple tasks is an incredibly complex mathematical routine. In fact even the most complicated robotic devices, with pattern recognition intelligence, have only 10^3 intelligence units, whereas the number of synaptic connections in the human brain is 10^{14}. Thus there is really no comparison at all between the intelligence of human beings and the potential of the most complicated robotic device. Yet we have the absurd situation where frantic efforts are being made to produce these robots, while we have 1 ¼ million people with infinitely greater intelligence wasting away in the dole queue.

In addition to the idea of alternative products on which we might work, of equal importance is that part of our corporate plan which deals with entirely different forms of work organisation. Modern industrial society is characterised by narrow fragmented tasks carried out at a frantic tempo. It is difficult for people not directly involved to realise just how frantic this is in certain parts of industry. We find, therefore, on the one hand that there is a lengthening dole queue, reflecting growing structural unemployment, while on the other those who remain in work find that they are subjected to an intolerable work tempo. I would quote, for example, the elements which make up an agreement between the unions and British Leyland (a nationalised concern) for the press workers on the Allegro car. The total allowance for rest periods was 32.4 minutes, made up as follows: trips to the lavatory, 1.62 minutes – not 1.6 or 1.7, but 1.62, computer precise – recovery from fatigue, 1.3 minutes; sitting down after standing too long, 65 seconds; for monotony, 32 seconds; and so on.

I say that a socialist economy cannot be based on a technology which is as brutal and depraved as that. Not only does it burn up human beings (in some car plants they reckon that human beings are 'burned up' in 10 years); it also burns up incredible energy as a result of being capital-intensive rather than labour-intensive. We are attempting in the corporate plan to challenge the assumptions about the manner in which modern production is carried out.

In conclusion, we believe that the product ranges we are

proposing would provide not only enough work for our own members, but for thousands of others, at British Leyland and elsewhere. They would provide us with the opportunity, not only to work, but to work in dignity at a civilised tempo and with a civilised relationship one to another at the point of production. But above all this, they would also provide us with the opportunity of using our skill and ability in the interests of the community, rather than to make profits for the few.

A CONCRETE PROPOSAL

These kinds of ideas have long been spoken about in the labour and trade union movement. Now we are presenting to that movement at large our corporate plan. It is a concrete proposal; it is not merely an aspiration or moral assertion. It is a feasible proposal backed by six volumes, each of 200 pages, of closely argued technical and economic data. We sincerely hope that the labour and trade union movement will now take this proposal up, and that similar proposals will be pursued at other plants up and down the country.

In presenting this document to the labour movement, we will watch very closely to see if those who have spoken about this kind of thing for so long are now willing to demonstrate their support for it by their actions inside the House of Commons, outside the House of Commons, and in factories up and down the country.

First published in:
Defence Cuts and Labour's Industrial Strategy
Labour CND Pamphlet, 1976

Science and Social Action

First published in *Science and Social Action*, 1978

Marxist critics of capitalist society have tended to concentrate, at least since the turn of this century, on the contradictions of distribution. This they have done at the expense of a thoroughgoing analysis of the contradictions of production within technologically advanced society. This imbalance can hardly be attributed to a one-sidedness on Marx's own part. Central to volume 1 of *Capital* is the nature of the labour process and a "critical analysis of capitalist production". In this, Marx demonstrates that with the accumulation of capital – the principal motivating force – the processes of production are incessantly transformed. For those who work whether by hand or brain, this transformation shows itself as a continuous technological change within the labour process of each branch of industry, and secondly, as dramatic redistributions of labour among occupations and industries.

That the overall development of production since then should accord so closely with Marx's analysis is a remarkable tribute to his work, bearing in mind the sparseness of occupations and industries then, compared with their proliferation today. Whether this marxist analysis will be equally consistent and valid when applied to the science-based industries which have emerged since the Second World War is now a matter of considerable discussion. With the integration of science into the "productive forces" this question is one of growing significance. In some large multinational corporations, fifty per cent or more of all those employed are scientific, technical or administrative "workers". This has begun to pose, in a very practical way, the relationships between science as at present practised and society.

Up to the mid 1960s, there hardly seemed any useful purpose in raising this question. At that time, there was hardly a chink in the Bernalian analysis of twentieth-century science. In this analysis, science, although it was integral to capitalism, was ultimately in contradiction with it. Capitalism, it was felt, continuously frustrated the potential of science for human good. Thus the problems thrown up by the application of science and technology were viewed simply as capitalism's misuse of their potential. The contradictions between science and capitalism were viewed as the inability of capitalism to invest adequately, to plan for science, or to provide a rational framework for its widespread application in the elimination of disease, poverty and toil.

The forces of production, and in particular science and technology, were viewed as ideologically neutral, and it was considered that the development of these forces was inherently positive and progressive. It was held that the more these productive forces – technology, science, human skill, knowledge and abundant "dead labour" (fixed capital) – had developed under capitalism, the easier the transition to socialism would be. Furthermore science is rational and could therefore be counterposed to irrationality and superstition.

Science had, after all, through the Galilean revolution, destroyed the earth-centred model of the universe, and through Darwin had made redundant earlier ideas of the creation of life and of humanity. Science viewed thus appeared as critical knowledge, liberating humanity from the bondage of superstition: a superstition which, elaborated into the system of religion, had acted as a key ideological prop of the outgoing social order.[1] The past few years have seen a growing questioning of this rather mechanistic interpretation of the marxian thesis. There is now a growing realisation that science has embodied within it many of the ideological assumptions of the society which has given rise to it. This in turn has resulted in a questioning of the neutrality of

Science and Social Action

science as it is at present practised in our society. The debate on this issue is likely to be one of major political significance. The question extends, far beyond that of scientific abuses, to deeper considerations of the nature of the scientific process itself. Science done within a particular social order reflects the norms and ideology of that social order. Science ceases to be seen as autonomous, but instead as part of an interacting system in which internalised ideological assumptions help to determine the actual experimental designs and theories of scientists themselves.[2]

Part of the importance of Stephen Bodington's present book is that it opens up some of these issues for discussion. For failure to deal with these questions will mean that the anti-science movement of the 1970s, whose antecedents lay in the anti-culture movement of the 1960s, will not develop beyond its initial and partly negative premise. According to this premise, science is viewed as evil, totalitarian and devoid of any attributes which might make it amenable to the "human spirit". This total rejection, now common amongst many young people, can, if properly handled, be elevated into a much more mature questioning of the fundamental nature of science and technology as practised in Western society.

Our Western scientific methodology is based on the natural sciences. Within this, relationships are mathematically quantifiable. There has been a tendency to suggest that if you cannot quantify something it really doesn't exist. This is not without its political significance; for if the mass of ordinary people are incapable of providing "scientific reasons" for their judgements (which are based on actual experience of the real world), ruling elites can then bludgeon their common sense into silence. Furthermore, attempts to use this narrow mathematically based science in the much more complex and indeterminate social sciences, and in political activity, can give rise to very serious distortions which flow inevitably from the abstracted nature of the scientific method.

It is significant that those working in the scientific field are themselves beginning to raise these questions. Thus Professor Silver has said:

"[There are risks] in the scientific method, which may abstract common features away from concrete reality in order to achieve clarity and systematisation of thought. However, within the domain of science itself, no adverse effects arise because the concepts, ideas and principles are all interrelated in a carefully structured matrix of mutually supporting definitions and interpretations of experimental observation. The trouble starts when the same method is applied to situations where the number and complexity of factors is so great that you cannot abstract without doing some damage, and without getting an erroneous result."[3]

Those working in the field of cybernetics have also expressed their concern about this misuse of "science":

"There is no doubt that a very important influence nowadays is a revised reductionism in the theory of cybernetics. It reduces processes and complex objectives to black boxes and dynamic control systems. Not only in the natural sciences, but also in the social sciences."[4]

None of this should imply the abandoning of a "scientific method". Rather we should understand that this method is merely complementary to, and should not override, experience, and "experience includes experience of self as a specifically and differentiatedly existing part of the universe of reality", as Stephen Bodington points out. Such a wider view of knowledge would free us from the dangers of "scientism", which, it has been suggested, may be nothing more than a Euro-American disease.[5] This view of knowledge implies a society which has a social structure capable of nurturing the coexistence of the subjective and the objective, tacit knowledge based on contact with the physical world, and abstracted scientific knowledge: in short, a society which would again link hand and brain, and permit people to be full human beings. This will mean challenging the fundamental assumptions of our present society, and indeed the assumptions of societies such as exist in the so-called socialist countries. One of the important factors now moulding the social forces and giving rise to such a challenge is the number of contradictions of science and technology experienced by an ever-increasing section of the population in our profit-orientated society. Indeed, the situation is now rapidly developing where scientists and technologists are being subjected in their work to many of the problems which skilled manual workers experienced at an earlier historical stage, when high capital equipment was introduced into their field of work.

The use of fixed capital (i.e. machinery and, latterly, computers) in the productive process marked a fundamental change in the mode of production. It cannot be viewed merely as an increase in the rate at which tools are employed to act on raw material. The hand tool was entirely animated by the worker, and the rate at which the commodity was produced – and the quality of it – depended (apart from the raw materials, market forces and supervision) on the strength, tenacity, dexterity and ingenuity of the worker. With fixed capital, that is the machine, it is quite the contrary, in that the method of work is transformed as regards its use value (material existence) into that form most suitable for fixed capital. The scientific knowledge which predetermines the speeds and feeds of the machine, and the sequential movements of its inanimate parts, the mathematics used in compiling the numerical control programme, do not exist in the consciousness of the operator; they are external, and act through the machine as an alien force. Thus science, as it manifests itself to the workers through fixed capital, although it is merely the accumulation of the knowledge and skill now appropriated, confronts them as an alien and hostile force, and further subordinates them to the machine. The nature of their activity, the movements of their limbs, the rate and sequence of those movements – all these are determined in quite minute detail by the "scientific" requirements of fixed capital. The objectivised labour in the form of fixed capital emerges in the productive process as a dominating force opposed to living labour. We shall see subsequently, when we examine concrete situations at the point of production, that fixed capital represents only the appropriation of *living* labour, but in its sophisticated forms (computer hardware and software)

The Search for Alternatives

appropriates the scientific and intellectual output of the white-collar workers whose very own intellects also oppose them as an alien force.

Therefore the more that workers put into the object of their labour, the less there remains of themselves. The welder at General Motors who takes a robotic welding device and guides its probes through the welding procedures of a car body is on the one hand building skill into the machine, and on the other hand deskilling himself. The accumulation of years of welding experience is absorbed by the robot's self-programming systems and will never be forgotten. Similarly, mathematicians working as stressmen in an aircraft company may design a software package for the stress analysis of airframe structures and suffer the same consequences in their jobs. In each case they have given part of themselves to the machine, and in doing so have conferred life on the object of their labour – but now this life no longer belongs to them but to the owners of the object.

Since the product of their labour does not belong to the workers, but to the owner of the means of production in whose service the work is done and for whose benefit the product of labour is produced, it necessarily follows that the object of the workers' labour confronts them as an alien and hostile force, since it is used in the interests of the owner of the means of production. Thus this "loss of self" of the worker is but a manifestation of the fundamental contradictions at the base of our society. It is a reflection of the antagonism between the interest of capital and labour, between the exploiter and the exploited. Therefore fixed capital, at this historical stage, is the embodiment of a contradiction, namely that the means which make possible the liberation of the workers from routine, soul-destroying or backbreaking tasks, are simultaneously the means of their own enslavement. It is therefore obvious that the major contradiction can only be resolved when a change in the ownership of the means of production takes place.

I have attempted to show elsewhere how this loss of self, or alienation, is now spreading rapidly in the wake of computerisation

Science and Social Action

into highly creative fields of intellectual work such as architecture, and scientific fields of work such as aerospace design.[6] Technological change of this kind means that the organic composition of capital is changed.[7] Processes become capital intensive rather than labour intensive, as workers by hand and brain are replaced by "dead labour". This is now giving rise to massive structural unemployment throughout most of the technologically advanced nations, and is a tendency which is going to continue. It does, however, mean that groups of "workers" such as scientists and technologists, who have never previously had experience of the degradation of the dole queue, are now feeling all the pangs of job insecurity[8] and are therefore much more likely to align themselves with those sections of the working class and critics of society who recognise the need for fundamental changes in the kind of priorities our society is setting itself.

There is, in addition, the disillusionment and the loss of job satisfaction which flows from the general tendency to reduce the level of qualification in all jobs.[9] Few manual workers any longer harbour illusions that science and advanced technology will liberate them from soul-destroying, backbreaking routine tasks and leave them free to engage in more creative work. This realisation flows from the sheer day-to-day experience of the dehumanised, alienat-ed tasks in modern production. This dehumanisation is due in large part to the fragmentation of work into narrow tasks, each minutely timed – the very essence of Taylorism; as its founder Frederick Taylor explained,

> "The workman is told minutely just what he is to do and how he is to do it, and any improvement he makes upon the orders given to him is fatal to success."

Initially, Taylorism on the shop floor actually increased the intellectual activity of staff in the offices. Taylor himself explained in his 1903 book *Shop Management* that his system was

> "aimed at establishing a clearcut and novel division of mental and

The Search for Alternatives

manual labour throughout the workshops. It is based upon the precise time and motion study of each worker's job in isolation, and relegates the entire mental parts of the tasks in hand to the managerial staff".

Seventy years of scientific management has seen the fragmentation of work grind through the spectrum of workshop activity, engulfing even the most creative and satisfying jobs, such as toolmaking. Throughout that period, most industrial laboratories, design offices and administrative centres were the centres of the conceptual, planning and administrative aspects of work. In those areas, one spur to output was the dedication to the task in hand, an interest in it, and the satisfaction of dealing with a job from start to finish.

The objective base for Taylorism was laid in the field of intellectual work when large numbers of scientific and technical workers were involved in the labour force as a whole. It was evident that eventually their intellectual activities would be divided into routine tasks and work study would be used to set precise times for its synchronisation with the rest of the "rational procedure".

In 1830 the father of the computer industry, Charles Babbage, in his *Economy of Machinery and Manufactures,* anticipated the effects of Taylorism in the intellectual field. He wrote:

"We have already mentioned what may, perhaps, appear paradoxical to some of our readers – that the division of labour can be applied with equal success to mental as well as mechanical operations, and that it ensures in both the same economy of time."

A paper in the journal *Work Study* (June 1974), entitled 'A Classification and Terminology of Mental Work', suggests that much "progress" has already been made in this direction. Having identified the hierarchy of physical work as "Job, Operation, Element, Therblig", it states:

"The first three of these are general concepts, i.e. they could apply equally to mental or physical work. The last term, the therblig, is specific to physical work. All elements of physical work consist of a small number of basic physical motions, first codified by Gilbreth (therblig = anagram of Gilbreth) and later amended by the American Society of Mechanical Engineers and the British Standard Glossary. The logical pattern would be complete if a smaller breakdown of elements into basic mental 'Motions' or yaks were available."

The paper discusses in detail how to classify input yaks, output yaks and processing yaks. It describes how each of these can be subdivided into basic mental operations. It goes so far as to draw a division between passive reception of visual signals (seeing) and active reception (looking), and passive reception of auditory signals (hearing) and active reception (listening). The paper implies that these techniques will be used in the more clerical aspects of mental work. However, it concludes:

"We have tried to show that mental work is a valid and practical field of application for *Work Study*, that basic mental motions exist and can be identified and classified in a meaningful way provided that one does not trespass too far into the more complex mental routines and processes. A set of basic mental motions has been described, identified, named, described and coded as a basis for future measurement research, leading to the compilation of standard times. There is a good prospect that such times would play a valuable part in *Work Study* projects."

Whether or not one regards this type of research as pseudoscientific, there can be little doubt about how it will be deployed. The employer of scientific, technical and administrative staff (including some forms of managerial staff) will see it as a powerful form of psychological intimidation to mould their intellectual work to the mental production line. It is perhaps a recognition of this tactical importance which prompted Howard C. Carlsson, a General Motors psychologist, to say, "The computer may be to middle management what the assembly line is to the hourly worker." Industrial experience already demonstrates that Taylorism is destined not just for those who use computers or are

The Search for Alternatives

in middle management, but for a range of intellectual "workers" including scientists and technologists.

It is true, as Stephen Bodington points out, that some intellectual workers have deliberately chosen to work in an academic environment in order to free themselves from the rat race, and the pressures which are typical of modern industry. It is also true, of course, that some of these academics, who are engaged in so-called "neutral science", are quite indifferent to the consequences their work has upon workers at the point of production. We now have the rather ironic development in which some of those people who at university helped to develop the scientific management production systems which have made work so grotequse for those on the shop floor may soon be the victims of their own repressive techniques. An article entitled 'College of Business Administration as a Production System' (which appeared in the *Academy of Management Journal,* vol. 17, no. 2) is symptomatic of a general tendency.

The terminology used to describe academic features and activities in the form of a factory model is strongly indicative of the underlying philosophy. Thus the recruitment of students is referred to as "material procurement", recruiting of a faculty as "resource planning and development", faculty research and study as "supplies procurement", instructional methods planning as "process planning", examinations and award of credits as "quality coutrol", instructor evaluation as "resource maintenance" and graduation as "delivery". The professors and the lecturers are of course "operators" and presumably, as on the shop floor, only the effective operators can be tolerated ("effective" for what and for whom, we may ask).

Those academics engaged in the physical and pure sciences will be pleased to learn that the important issues of efficiency and optimisation will not be left to the subjective ramblings of the sociologist or the tainted ideology of the political economist. The full analytical power and neutrality of real science, and the penetrating logic of mathematical method, have been brought to bear. They will undoubtedly produce a completely "objective" solution. I have referred elsewhere to the use of the well-known FrankWolfe algorithm by Geoffrion, Dyer and Freiberg, who define six criteria for optimising efficiency in the university, and then proceed to show how it can be determined.[10]

Programmes to increase faculty productivity are already spreading rapidly in the United States: recent grants indicate the potential scale. For example, a centre for professional development was set up in the California State University and College System with a grant of $341,261 from the Fund for the Improvement of Postsecondary Education in Washington. A comprehensive faculty activity analysis questionnaire has been prepared and developed by the University of Washington. The actual percentage time devoted to each faculty activity is requested. Each activity is specified very precisely, as it might be in a factory. Even one's own reading activities are included. There are some academics who hope that in projects of this kind, educational requirements will outweigh mere productivity considerations. However, a growing sector feels the outcome will be a smaller faculty teaching more classes and more students. This has already happened at the City University of New York, where seven hundred faculty members were sacked *(New Scientist,* 22 April 1976).

Increased "productivity", however, could have consequences much more widespread and subtle than the obvious ones of increased work tempo, loss of control, job insecurity and even redundancy. The impact it will have on the creativity of those involved is likely to be significant, for central to all optimisation procedures of this kind is the notion of specific goal objectives. A vivid example of the need to avoid such an over-constrained work environment was the design recently of EMI's computer-controlled brain and body scanner. In his evidence to the Select Committee on Science and Technology, Dr John Powell, EMI's Managing Director pointed out that the scanner was developed using unallocated funds as a by-product of work on optical character recognition. He stated in evidence that had its inventor "been constrained to follow a set objective on contract research funded by an operating division, he might have just produced

The Search for Alternatives

another optical character recognition machine."

Thus even the elitist right of the scientific worker or the researcher to give vent to his or her creativity will now be increasingly curbed by the system as it seeks more and more to control human behaviour in all its aspects. This is part of the general process by which the small elite who control society gain complete control over all those who work, whether by hand or brain, and use scientific management and notions of efficiency as a vehicle for doing so. It will be seen, then, that the organisation of work, and the means of designing not only jobs but also the machines and computers necessary to perform them, embody profound ideological assumptions. By regarding science and technology as neutral, we have

> "failed to recognise as anti-human, and consequently to oppose the effects of values built into the apparatus, instruments and machines of their capitalist technological system. So machines have played the part of a Trojan Horse in their relation to the labour movement. Productivity becomes more important than fraternity. Discipline outweighs freedom. The product is in fact more important than the producer, even in countries struggling for socialism."[11]

It has been suggested[12] that by ignoring these considerations the Soviet Union was laying the basis for the present situation in which it would be hard to argue that a worker there enjoys the sense of fulfilment through his or her work envisaged by the early marxists. It may well be that instead of developing entirely different forms of science and technology, and merely trying to adapt those developed in the capitalist societies to its own society, the Soviet Union made a profound error. Indeed the development in that country must find part of its origins in the attitude of Lenin to Taylorism, which he said,

> "Like all capitalist progress, is a combination of the refined brutality of bourgeois exploitation, and a number of the greatest scientific achievements in the field of analysing mechanical motions during work, the elimination of superfluous and awkward motions, the

Science and Social Action

elaboration of the correct methods of work, the introduction of the best system of accounting and control etc. The Soviet Republic must at all costs adapt all that is valuable in the achievement of science and technology in this field. The possibility of building socialism depends exactly on our success in combining Soviet power and the Soviet organisation of industry with the up-to-date achievements of capitalism we must organise in Russia the study and teaching of the Taylor system and systematically try it out and adapt it to our ends."[13]

Socialism, if it is to mean anything, must mean more freedom rather than less. If a worker is constrained through Taylorism at the point of production, it is inconceivable that he or she will develop the self-confidence and the range of skills, abilities and talents which will make it possible to play a vigorous and creative part in society as a whole.

So it would appear that in the technologically advanced nations there are now beginning to emerge a range of contradictions which will demand a radical re-examination of how we use science and technology, and how knowledge should be applied in society to extend human freedom and development.

It is precisely problems and contradictions of this kind which have caused more and more discussion about these issues: first in France during the late 1960s, and currently in the United Kingdom. Issues of this kind were central to the development of a Corporate Plan by workers at Lucas Aerospace, the British-based multinational. There, scientists and technologists realised that they had a direct community of interest with those on the shop floor, and they established a Combine Committee which, in the words of its Secretary Ernie Scarbrow, "Combined on the one hand the analytical power of the scientist and technologist, with the equally (or more) important common sense of those on the shop floor on the other". This, he pointed out, provided a unique and volatile combination. It is the kind of combination that one hopes will be witnessed in more and more large plants. In Lucas Aerospace the highest-level technologists and workers on the shop floor were together subjected to the degradation of structural unemployment. But not only that: they saw also the gap between what their

The Search for Alternatives

ingenuity, skill and ability could provide and what it was actually providing in our society. They were engaged during the day on designing, developing and building equipment to be used in an aerospace system as sophisticated as Concorde and on their journey home were passing through (or living in) communities where old-age pensioners were dying of hypothermia. They were also subjected to the gradual deskilling of their functions and a complete loss of control, since Lucas Aerospace, like other large multinationals denied them the right of freedom of expression in the course of their daily work.

So they put forward a six-volume Corporate Plan, suggesting a whole range of products which society needs, and which they had the skills and abilities to produce. They questioned the underlying assumptions of the market economy which prevented them doing that, and began to consider products for their use value rather than their exchange value. They also opened a debate in which, as workers by hand and brain, they thought of themselves in their dual role in society – that is, both as producers and as consumers. Lastly, and probably most importantly, they proposed an entire reorganisation of the labour process, in which hand and brain could again be linked, and workers could give full vent to their creativity. For the Lucas workers believe "that society's greatest asset is the skill, the ability, the ingenuity and the energy of its ordinary people."

The concepts underlying their Corporate Plan are now being discussed by shop stewards representing all levels of workers from scientists to semi-skilled workers at companies like Vickers, Parsons, Rolls-Royce and elsewhere. This is seen not just in organisational terms, but as questioning the nature of the equipment design itself. To this end the Lucas workers are proposing a range of equipment which would reverse the historical tendency objectivising human knowledge and confronting the worker with an alien and opposite force as described earlier. The equipment is known as "telechiric" ("hands at a distance"), in which the human being would be in control, real time, all the time, and the system would merely mimic human activity but not objectivise it. Thus the producer would dominate production, and the skill and ingenuity of the worker by hand and brain would be central to the activity and would grow and develop.

This development can begin to lay the basis, in conjunction with wider political action, for the transformation of society away from its present exploitative hierarchical forms, to a new form of society which, as the founder of cybernetics Norbert Wiener once said, differs from those propounded by many successful businessmen and politicians. In 1950, explaining his views on cybernetics, he wrote:

> "I should like to dedicate this book to the protest against this inhumane use of human beings. It is easier to set in motion a galley or factory in which human beings are used to a minor part of their full capacity only, rather than to create a world in which these human beings may fully develop. Those striving for power believe that a mechanised concept of human beings constitutes a simple way of realising their aspirations to power. I maintain that this easy way to power not only destroys all ethical values in human beings, but also are very slight aspirations for the continued existence of mankind."[14]

Stephen Bodington raises these and many other issues. There may be those who will disagree with some of the ideas propounded, but that should not prevent them from entering into an honest discussion around the issues which he raises. For it is only through a really active political exchange of ideas, and by relating these to our direct experience so that theory and practice are constructively linked, that a solution will come.

It is to be hoped also that this debate will not be limited to an introverted soul-searching of the scientific community. These matters can only be resolved when they involve large numbers of the entire population in a profound discussion about them. The experience at Lucas shows that workers by hand and brain, who are capable of designing and building all the real wealth we see about us, are equally able to discuss these issues even though they may not use the same terminology as intellectuals.

The Search for Alternatives

Notes

1. Hilary and Steven Rose, "The Incorporation of Science'', in *The Political Economy of Science,* Macmillan, London, 1976.

2. Hilary and Steven Rose, in W. Fuller (ed.), *The Social Impact of Modern Biology,* Routledge & Kegan Paul, London, 1971.

3. R. S. Silver, "The Misuse of Science", in *New Scientist,* vol. 66, no. 956, 5 June 1975.

4. D. Henning, "Bericht 74-09", Berlin Technical University, Fachbereich 20, January 1974.

5. J. Needham, "History and Human Values", in H. and S. Rose (ed.), *The Radicalisation of Science,* Macmillan, London, 1976.

6. Mike Cooley, *Computer Aided Design,* AUEW, London, 1972.

7. Mike Cooley, "Contradictions of Science and Technology", in H. and S. Rose (ed.), *The Political Economy of Science,* Macmillan, London, 1976.

8. Mike Cooley, "Taylor in the Office", in R. Ottoway (ed.), *Humanising the Workplace,* Croom Helm, London, 1977.

9. Harry Braverman, *Labour and Monopoly Capital,* Monthly Review Press, New York, 1974.

10. Mike Cooley, "The University as a Factory", in *New Scientist,* vol. 70, no. 1006, 24 June 1976.

11. Robert Jungk, *Qualitiit des Lebens,* EVA, Cologne, 1973.

12. Braverman, *op. cit.*

13. V. I. Lenin, "The Immediate Tasks of the Soviet Government" (1918), in *Collected Works,* vol. 27, Moscow, 1965.

14. Norbert Wiener, *The Human Use of Human Beings* (1960), p. 16.

First published in:

Science and Social Action
By Stephen Bodington, Allison & Busby, 1978

THE LUCAS AEROSPACE CORPORATE PLAN:
AN INTERVIEW WITH MIKE COOLEY

First published in *Revolutionary Socialism* magazine, 1980

In the mid-seventies the Lucas Aerospace Combine Shop Stewards Committee launched a campaign round its own workers' plan, which sought to act as a tool for mobilising the entire workforce on the need to fight to produce socially useful products. The struggle of the Lucas Aerospace workers has inspired others in Britain and other countries to adopt a similar strategy and also provoked controversy among the left. Mike Cooley, who is a member of the Lucas Aerospace Combine, describes the ideas behind the plan and some of the many problems involved. The interviewer is David Harding.

DH. I'd like to start by asking you to say something briefly about the nature and history both of the Lucas Aerospace Combine Committee and the development of the Alternative Corporate Plan within Lucas Aerospace.

MC. Well I feel the Plan can only be properly understood against the background of the restructuring of British industry at the end of the 1960s. The then Labour government suggested that it would be in the national interest and in the interests of individual corporations if they were to be enlarged and then nationalised so that you didn't get duplication of effort and research, and so it gave millions of the taxpayers' money through the Industrial Reorganization Corporation to set up vast corporations like GEC.

The Search for Alternatives

Weinstock, the managing director of GEC, brought together something like 260,000 workers and within 18 months he had reduced that to 190,000. Weinstock was able to do that because he succeeded in setting one factory against the other and one group of workers against the other. So it could be said that we learnt through the negative example of what happened at GEC.

We could also see the inability of the existing trade union structures to cope with a multi-union, multi-site, multi-national company. They were divided into districts, and in crafts, and management played ducks and drakes with them, set one against the other. So we started to build our Combine Committee so that this could not be done to us. There were very important developments like getting our own newspaper so that each site knew what the other one was doing and so on but we were not able to build a Combine quickly enough to prevent a very serious defeat in 1971 in the Willesden factory in London. We'd occupied it for about six weeks and the morale of the workforce had begun to decline because we were campaigning for the same old right to work, on the same old products in the same old way and younger workers could see no future in that and they could get jobs elsewhere. Some of the older workers had seen that in all the other battles against unemployment the workforces were defeated. So although we'd occupied the place and were very militant, in a traditional sense, we were defeated because on the sixth weekend the morale had declined so much that the factory wasn't occupied and the company brought in a demolition group and tore the roof off. Now, out of that negative experience a tremendous discussion started about what we should be doing and somebody asked a very simple question. They said why can't we use all the skill and ability we've got to meet all the social needs we can see about us? Those social needs are absolutely glaring in an advanced industry like ours. We're making the generating equipment for Concorde for example yet our manual workers live in communities where old age pensioners are dying of hypothermia. So the gap between those two is enormous and it was therefore firstly necessary to set up a trade union organization which could be a vehicle for the

The Lucas Aerospace Corporate Plan

Corporate Plan and secondly to have this concept of linking people's roles as producers and as consumers.

DH. Can we look at progress you have made in the implementation of the Corporate Plan since 1975 and at some of the specific struggles that you have had, against redundancies, in some of the Lucas Aerospace plants. First of all how has the Lucas management reacted to proposals in the Plan and attempts to implement them'?

MC. Well I think they reacted at about three levels. Firstly, they didn't really believe that the thing was going to take off. They really believed that it was just a few activists who had these concerns about technology being used in an appropriate or socially useful way and that it didn't reflect the reality that most industrial workers experienced, and in that they were absolutely mistaken. Second, having failed to create circumstances where the thing died off itself the company then set up an alternative Combine Committee. They got a group of manual workers who were used as stooges of the management to set up an alternative combine to try and break our one up. Third, they tried to involved the official trade union movement against us by saying we were an unofficial body and that they couldn't negotiate with us because the full time officials wouldn't allow it. At the same time they were saying to the full time officials, if this kind of thing gets going all the links we have with you will be broken, the whole thing will be chaotic and so on.

At a direct industrial relations level they have tried to sack many of the leading people of the Combine. They even arranged a situation where they said that one of the security people at the Wolverhampton plant had been assaulted by the Convenor of Shop Stewards. They were going to set up a trial where they would take his job away from him. More recently they've tried to sack Ernie Scarbrow.

DH. What have you achieved?

MC. The specific things we have achieved are first, we've

The Search for Alternatives

demonstrated beyond any doubt the ability of so-called ordinary workers to decide what products they should make, how they should make them and in whose interests they should be made. And so much so that they've seen through the whole myth of hierarchical management. As one of the Burnley workers put it: we've discovered that management is not a skill or a craft or a profession but a command relationship, a bad habit inherited from the army and the church. A very high level of consciousness coming through. And I think that in itself would have made the Corporate Plan worthwhile. But in a much more profound sense it means that we have actually built prototypes of the products that we are talking about. There are now products in existence, true they are mainly as prototypes, which demonstrate that you can so use technology to devise products for their use value and not just for their exchange value: products that conserve energy, materials, and enhance skill. The most important of these, I think, is the road/rail vehicle. We're now building a coach body onto it and plan to drive the vehicle round to the different towns where we've got factories, park it in the market place, or outside the town hall, or whatever, and have inside it a series of slides, video-tapes, posters where people can come in and discuss with us what our kind of technology could mean to their community.

DH. In the face of the reaction from management and from the official trade union structure what is your current strategy to push forward the implementation of the Corporate Plan? How do you propose to overcome this blockage?

MC. Well the blockage became more complicated two years ago when the company tried to close down two factories, one in Liverpool and one in Bradford. As a result of all the publicity the company agreed to intervention by the Labour government to create a tripartite meeting with the unions, the management, and the government. In our view that was really disastrous for us because that took the whole campaign and the activity out of our hands and placed it in the hands of the bureaucrats. The end result was that the Labour government gave Lucas £8 million of the

The Lucas Aerospace Corporate Plan

taxpayers' money to build a new factory in the Liverpool area. So our main strategy now, having gone through all the political hoops that everybody told us one should go through, is to concentrate on big combine committees in other groups throughout the country, concentrate on the grass roots which experiences the problem of structural unemployment day by day.

DH. That is taking the struggle beyond the specific experience of Lucas into other companies such as Vickers, Thorn etc?

MC. Yes, because we believe that the Corporate Plan strategy, as a strategy, is appropriate to any industry or any community; if they simply look at their own resources, facilities and skills and they link them with the needs of that community.

DH. Are you also saying perhaps that the experience of Lucas shows that at the level of one individual plant it is very difficult to implement an alternative plan, given a hostile management, and that you need a more global political approach'?

MC. Yes. Well, firstly I think it is inevitable that the management structure will be hostile to us because the value systems are entirely different. But I agree completely with you. In our view as we said right on page one of the Corporate Plan there can be no islands of social responsibility in a sea of depravity. We've done what we did to demonstrate the ability of people to do it and any other group of workers in our view could have done it. So we think it is absolutely central that it be spread out and we've never felt we could do it in isolation.

DH. I want to ask two questions about the ability of the Combine Committee to maintain the support of the base over the years since 1975. Given the difficulties since the plan was published was it not difficult to sustain the dynamic of discussion, of organization, over the period? And at the more fundamental level of redundancy struggle you mentioned the closures in several plants. Has the Combine Committe been able to avoid the sort of closures and redundancies management wanted to impose?

The Search for Alternatives

MC. I think the dynamic has been maintained for a number of reasons. One is that the problems we identified as the growing problems in technologically advanced society have got signiticantly worse since we produced the Corporate Plan, and the need for this sort of thing seems to us to be greater now. The second thing is that the interest outside Lucas, including international interest in the Plan, has really caused Lucas workers to realise that they are handling something which is incredibly important. All the interest, all the other people writing about it, who were beginning to take up similar ideas including in the United States, in Detroit and elsewhere, has reinforced the need to try and increase the dynamic and maintain the involvement. But it has at times been difficult and frankly it becomes very difficult whenever there isn't a direct threat of unemployment. I'd have to be absolutely frank and say that the lowest common denominator has always been the primitive fear of unemployment and the degradation that flows from it. But our analysis that there would be continued attacks on the workforce as they tried to cut back production in the United Kingdom and expand abroad has been proven to be correct. Every so often they try to close some factory which restimulates the whole thing again.

DH. And in the specific example of the Liverpool plant, were you successful in opposing redundancies?

MC. Since we've produced the Corporate Plan we've been able to prevent the company sacking even one Lucas worker directly. It's probably the only multinational, multi-union Company in the country where that is true. Several times they've tried to close plants and we've been able to prevent them because we have a level of consciousness now amongst the workforce which will do that. Though we are not able to cope with natural wastage, we have prevented them from closing either the Bradford factory or the Liverpool factory.

DH. Can we look at the relationship between the Combine Committee and the official trade union structure in Lucas. The formation of the Combine Committe represented the development of a parallel structure of representation in Lucas, a more embracing structure than that of the official unions. What was their reaction to the Combine Committee's formation and has that reaction changed, given that many unions now, at a national level, support the general idea of Corporate Plans?

MC. Well I think that the reaction of the trade union bureaucracy to the Combine Committee is almost identical to the reaction of the trade union movement to the shop stewards' movement at the turn of the century. The shop stewards' movement represented an objective need in industry which the trade union movement was not catering for. Once the workers themselves at the point of production began to meet that need they were met initally with hostility. Then the attitude seemed to be to try to incorporate them and in many ways they're now institutionalised within the trade union movement. I think that's what we're witnessing with the Corporate Plan. From the onset the TUC said it agreed with the Corporate Plan and with this idea of socially useful work. Indeed they produced a film about it as part of their shop stewards' training programme. But the individual unions perceived it as a threat to their authority and many of them were openly hostile and that hostility wasn't limited to the right wing unions. It was strong from some of the left wing unions who also have notions of narrow hierarchical control. Quite early on the T&G accepted the Corporate Plan as its policy. It issued a booklet in relation to defence conversion where the majority of the book is based on the Corporate Plan. What we notice now is that some of the individual unions are setting up what they call 'Combine Committees' which means that they have an umbrella ogranization which covers members in their union only in the plants throughout the whole of the country. This in our view has the danger that it will fragment at the combine level the situation which is bad enough at a plant level. In fact in some ways it's making things worse.

DH. What is the relation of plant shop stewards to the Combine Committee over particular defensive struggles?

The Search for Alternatives

MC. The Combine Committee provides them with a sort of framework in which everything which goes on at all the other plants is immediately conveyed to them. When they get into difficulties support is mobilized even on issues like wages. If a group is locked-out on wages then we prevent work coming in from that factory and so on. So it's an enormously supportive organization for their own individual struggles and it has deliberately ensured that it has no powers other than that of persuasion. It can only recommend things to each site and therefore it's got to be very sensitively tuned to the requirements of individual shop stewards' committees on each site who should then be sensitive to the requirements of people on the shop floor. This is unlike the big combines in the 60s which tried to impose policies on the shop stewards throughout the sites and got considerable hostility from shop stewards in individual factories. To a certain extent I think we've overcome that, although I freely admit that some groups of shop stewards were turned against us partly on a manual versus staff basis by the company – the point I made earlier. The attitude of the full time officials to the Combine really differed enormously from union to union. The T&G officials will do everything possible for us. One of the national officers is now on the advisory centre we set up at North East London Polytechnic (CAITS – the Centre for Alternative industrial and Technical Systems). On the other hand some of the left wing unions and also some of the right-wing ones which organise white-collar workers are quite scathing about it because they do see it as somehow undermining their authority. In my view I think it should be regarded as a logical extension of the trade union movement as it stands.

DH. Can I ask a clarifying question here. We've talked about the relation at plant level between the Combine Committee and the official unions. How is that relationship structured?

MC. In every plant there should be a joint shop stewards' committee which represents all the unions on that site. And that shop stewards' committee should send two directly elected

The Lucas Aerospace Corporate Plan

representatives to each quarterly meeting of the Combine. In addition to the two representatives it is encouraged to send as many observers as possible because we believe that small elites are the incubators of corruption. Communication is maintained by the paper, by regular services which are provided like on wage bargaining, on new technology, on health and safety issues, on pensions and so on.

DH. Let's shift to a different area. The Lucas Aerospace Corporate Plan focuses its attention on a particular type of product – what you have called socially useful products. To what extent though, within the Plan, do you look at and attempt to develop a strategy for, first of all, control of the production process? And secondly to what extent have you looked at the production process in terms of new technology currently being introduced and the wide implications of that technology for the nature of the labour process and for future employment?

MC. In the original Corporate Plan we made the point about the need to link hand and brain, the idea being to release all the tacit knowledge and creativity of people on the shop floor which modern Tayloristic production techniques destroy. Now we haven't been able to force a position where any of the Plan's products is mass produced in Lucas or produced in any quantities. Heat pumps have been produced only as prototypes. So thefore, there hasn't been the opportunity to test out in practice a different labour process, much as we would like to do that. We have operated a different labour process both in drawing up the Corporate Plan itself and also in making the prototypes. For example, the road/rail vehicle which is quite a complex mechanism. No ratified mathematical analysis was done for that at all. We used the common sense, the experience and the tacit knowledge of workers by asking them what size they thought an axle should be and we made the axle that size and in fact it worked because these people have spent a lifetime making, bending and twisting axles. We were able to democratize the whole decision making process, within the product design, the product planning

The Search for Alternatives

and the product manufacture. So at a very small embryonic level when we were producing the prototype, we changed the nature of the labour process. Now we're putting forward a proposal to the company at the moment, in a collective bargaining framework for a new factory in the Liverpool area to produce heat pumps. We've got them running as prototypes and we have involved architectural students and others in so laying out the factory as not to reproduce the kind of hierarchy and control that exists in modern industries – to get real integration of workers by hand and by brain.

The whole question you raise has become much more significant in the last six months where the company is attempting to introduce significant pieces of new technology, computer aided design techniques, electro-chemical machining, numerically controlled machines, and so on. At the moment we're just starting to draw up a Corporate Plan on new technology where we're going to go through this same process again of involving workers in all the factories in a tremendous discussion, getting their ideas and so on. And in the meantime we're imposing a moratorium on all new technology. It yet remains to be seen how successful we'll be in doing that but at least we're raising the issue that workers should not have the new technology in until they've had a chance to analyse the consequences and begin to put forward a new proposal. I might add that in doing that we're only reflecting what the TUC in Australia has done. It's put a moratorium on all new technology for five years until it can examine the consequences.

DH. Have you at any time during your attempts to implement the Plan presented Lucas Aerospace management with an integrated formula which includes not only the 'socially useful' products but also aspects of the production process, the question of control over it?

MC. No, we haven't done that yet. We've fought initially on the basis of the socially useful products. That would be a significant breakthrough. If we could then get that we would do the second stage. But of course in our day to day working, on the conventional products, we exercise a significant amount of control over the pace

The Lucas Aerospace Corporate Plan

at which we produce, what the rates are, and so on.

DH. Can you give us an idea – if we look at examples of socially useful products such as the heat pump, the road/rail vehicle that you've mentioned – of the form of production process you would be arguing for in the implementation of that aspect of the Corporate Plan.

MC. Really, I think we regard management as we now know it as moribund and superfluous, just a command relationship, and what we would really like to see is a self-managed form of factory in the way in which we self-managed the whole production of the Corporate Plan itself or the way in which we self-manage the Combine Committee with all its finances, its newspaper, and everything else. We feel that it would be wrong to leave the final decisions on what parts are produced in factories and how they're produced to those locked into the factory situation. We think it's vitally important that they should be reflecting the needs of the communities in which they're based. Rut we're by no means clear how that element is going to be brought in. We've attempted quite unsuccessfully now, on five occasions, to get discussions going in communities about how factories should be run and how they could be involved in it.

DH. What you've just said implies a radical challenge not just to the 'function of management' but to the whole structure of social relations within the company, the whole question of who controls what in a capitalist framework. Does that not produce a fundamental reaction from a company such as Lucas Aerospace, from any company enmeshed in capitalist relations of production?

MC. Yes, sure. It's been absolutely hostile. There is one development going on when one talks about the reaction of management. It would be true to say that at the level of management which represents finance capital, that level has been absolutely hostile. These are the people deciding whether they will invest in Brazil or South Korea or elsewhere. But at the level of management which represents industrial capital, people who like

The Search for Alternatives

to see things made and produced and so on, as distinct from fiddling with paper money, we found a very receptive audience amongst those kind of people many of whom incidentally would lose their job if the factories were closed down in any case. So I wouldn't want to generalize too much about management as a whole, I'd have to draw the distinction between technical management and division managers, often closely linked into communities, where the kids are at school, where the factories are and who would themselves be in serious difficulties if there were closures.

DH. To what extent do you feel the need to look for a new strategy now around the Corporate Plan and its implementation? You've talked earlier about having gone through all the hoops, about the move by the company now to introduce a whole wave of new technology with serious consequences for both jobs and the labour process, and you've now outlined above the radically different nature of the production process you would ideally like to see. How do you then see your strategy in the next year?

MC. One avenue that some people see is identifying some products and then campaigning for new factories to be built to produce them. That would be one area. I don't think that is going to take us very far because the company will try to exercise ownership over the products and so on. I don't see any point in being involved with the government much longer, either the present one or the one we had previously. At present, there is brewing a major confrontation between workers and Lucas management over on whose terms new technology will be introduced.

DH. That will be a defensive struggle but one with links to the approach to production processes which is in the Corporate Plan?

MC. Yes, and it will be offensive to the extent that we've got very clear views about how new technology might be used. For example, the telechiric devices which were mentioned in the original Corporate Plan, devices that mimic in real time the motions of a worker going through a skilled labour process but do not objectivize the human skill or diminish the human being. So we won't just be saying you can't have this in the new technology unless you pay so much. We don't see it in narrow economic terms. We will be proposing ourselves alternative means of production which we would like to see introduced. So it will, as you say, on the one hand be defensive but on the other hand we will be also attacking from the other side in proposing our own forms of new technology. But I don't think we will be able to do much more in isolation. We can only move forward if other combines develop in other companies.

DH. Many of the products you've put forward in the context of the Lucas Corporate Plan relate to areas such as health and public transport. One of the aims of the Combine Committee is to break with a narrow economistic view of the union struggle or with its location in particular plants. What has been your experience in working with, and organizing with, workers in other sectors?

MC. In the case of designing factories like the heat pump factory we've been able to work with architects and architectural students and so on. But I think the most fruitful field so far has been with the public service unions and with the health service unions. We've had meetings with a medical panel we've set up where doctors, nurses, auxiliary workers in hospitals, all work together in an integrated team where they can identify requirements in the health services and we can very often quickly find technological solutions to those. The only barrier is an economic one or a political one or something else. So that has worked fairly well but it's still at a very embryonic stage because unfortunately there is still an enormous division between people as producers, that is their life in factories and their life in the communities.

In the case of the public services unions, they have been inviting us to their weekend schools, we've had a whole series of speakers at conferences they've organised to talk about ways of fighting the cut-backs and linking in directly. But again that is still at a very embryonic stage: the Corporate Plan has only been going for about

The Search for Alternatives

four years in its present form and it takes a long, long time to change political attitudes and organizational ways of behaving, particularly with traditional hierarchical ways of doing things that you often get in Labour or left movements. We've been trying to form new organizations as we've made new links; it's inevitably been a fairly slow process.

DH. Do you see, in the future, yourselves linking in with say health workers in a struggle over implementation of Corporate Plans on Lucas? How would you see that sort of joint struggle operating?

MC. We hope that we can make far greater use and far greater involvement with trades councils. In our view that's a completely underestimated form of trade union organization so we're putting a lot of emphasis on that at the moment. And we would like to see a situation where, for example, if a hospital is being closed down for lack of equipment, the health workers in that hospital would come to local factories and say 'this is the hospital where you and your family will have to go if they're ill,' and that the local factories should be involved in direct action with them, including strike action, to maintain those hospitals and if necessary appropriating part of their time and maybe equipment in those factories, waste materials or whatever, to produce machines or equipment which the hospital workers need.

DH. Do you have any experience to date, in Lucas factories where the Combine Committee is active, of workers taking action outside the factory in these wider struggles?

MC. Oh yes. There was a meeting in Burnley, a town hall meeting, where 300 people came along from the community to define community needs and so on. There were long discussions about that and the Lucas workers have been helping them for example to improve housing, conserve energy and materials. In some of the local struggles around hospitals our shop stewards have played a leading role with nurses and others in their area. So all that is developing, though quite slowly.

The Lucas Aerospace Corporate Plan

DH. Do you see the need to develop structures that go beyond those that exist at this moment, such as the trades councils you mentioned, to meet these wider struggles? It seems to me that historically, and I'm not sure that this is necessarily so, trades councils have always been very wary of taking an active role in cross union struggles. Certainly as they are constituted at the moment trades councils have shown themselves very weak precisely in the sort of area of struggle you've been talking about.

MC. I think in some ways that's a reflection of attitudes within the individual unions which make up the trades councils. Unions are very very possessive of their own patch and they are deeply ridden with economism. And that economism grew stronger at a time when Britain was a great imperialist nation, where the metropolitan working class could probably screw a bit more out of the metropolitan ruling class – although I'm not suggesting that the British working class were ever 'living it up'. But they didn't have to be as political as the workforce in other nations. Also I think in many ways the political dynamism was sapped by social democratic ideas through the Labour Party and elsewhere or if it was encouraged it was often by small left wing groups, the CP or those others who behaved in a very, very arrogant way towards the working class; instructing them what to do, telling them how to go about it and so on.

So I do see the need for new structures. I'm by no means clear what type of structures they should be. Clearly they will have to be political to transcend the narrow economism of the trade unions but I hope that they can avoid all these mistakes that were made in the past, these terrible mistakes about the role of leadership and so on. We've seen in developing the Corporate Plan and all that has flowed from that the importance of leadership being more catalytic and enabling, where you keep your most dynamic elements at the base rather than sending them up into the superstructure; where the base is always correcting and purifying the superstructure, rather than a small elite taking things over at the top. So yes, I see the need for new structures but if it is to be a

The Search for Alternatives

political organisation it must involve masses of people and its political ideas should be put over in a catalytic way rather than an authoritarian way.

DH. I want to look at a final area of discussion. We now have a new Conservative government whose industrial policy is radical in a very right-wing sense, and could prove to be one of the most reactionary we've seen for a long time. To me it would seem that the climate created by this government and its policies is a very negative one for implementation of Corporate Plans in a factory faced with closure or threat of redundancies. Two or three years ago under the Labour government there was at least always that possibility of state intervention to give a certain room for manoeuvre – time and money for restructuring, reorganizing, to produce new products, within a Plan structure. In the present political situation the state clearly has no intention of offering that sort of room for manoeuvre. Does that not make for a much bleaker outlook for the Corporate Plan strategy in the next few years?

Can I ask you also more generally about what you see as the basic relationship between the Corporate Plan movement and the state. Do you see it, if not as a necessity, than at least an advantage, something to work for, to have a state structure which operates in a reformist way, and will intervene directly, financially, to offer a certain room for manoeuvre to firms in crisis that could facilitate implementation of Corporate Plans? You said earlier that you had exhausted governments, wanted nothing more to do with them. Would you still hold to that if, for example, there were a left reformist government in power in England?

MC. Well firstly, as far as the present government is concerned certainly, outwardly, it's very aggressive towards the trade union movement. It's still difficult to know to what extent they differ in any substantial way from the last government in relation to their attitude to big companies. The other government had certainly a softer approach when there were redundancies and so on, but it

The Lucas Aerospace Corporate Plan

was still governed by market forces, and our experience under the Labour government was they never gave us any direct support. There was a lot of sympathy but it always ended short of actual help. There were pious statements in the House of Commons and so on, and I detect that the Thatcher government is still in many ways planning industry in an indicative sort of way. They haven't got rid of the National Enterprise Board in spite of all the political assertions that they would do so. They're still making massive funds available for research and development, so they're not leaving it to free market forces. And the whole tendency internationally is to have the state intervening. Now I can see the need for a state that is sympathetic but the real danger is that that would be the kind of state that existed in Sweden say for 40 years where workers, in my view, are absolutely passive, where when there's a problem they always look for the next representative to deal with it for them, where they're incorporated into the thinking of the government and they always see the solutions as within the framework of government. I think we're looking for something far more democratic and outgoing although we recognize that there must be a state which reflects that. But we don't for example believe that if large sections of British industry were nationalized in the way that the Bennites and others suggest that would necessarily make it more socialistic. We believe very much in what James Connolly once said that if nationalization alone meant socialization, the hangman would be a socialist because the hangman is nationalised as well. We need a very different state structure to make that sort of thing possible and I don't see a future Labour government doing that very well. For example we have always recognized the need for the state to intervene in medical care and we think it was a very, very important development in Britain that a Labour government took medical care to a large extent outside the framework of market forces. This was profoundly important. We feel there are whole other areas, including parts of manufacturing industry where the main concern should be the use value of products rather than their exchange value and that was what was significant about the National Health Service.

The Search for Alternatives

So certainly we support that kind of development but in practice the dangers have been that in implying that they were going to do that with large sections of British industry and restructuring it the last Labour government succeeded in conning the trade union movement and its leadership in particular into accepting 1.5 million people out of work. Now had that been done under a Conservative government I'm sure there would have been an outrage. I can recall 15 years ago people saying that there would be a confrontation between the government and the unions if we ever got half a million people out of work. We've now got 1.5 million and it was the Labour government that was able to con people into that. So whilst I'm not taking a sectarian view and saying that we don't welcome intervention, I'm trying to make the point that intervention by the state itself, even under a Labour government, won't get the kind of thing we want. It's going to require radically different policies and in my view those policies will only ever be fought through and sustained and supported if masses of people at the factory level and the community level are involved in ensuring that they're implemented in their interests. And that they're part of the process, rather than a small elite, even a left wing elite in the Labour government, doing it by proxy on their behalf.

First published:
Revolutionary Socialism
Big Flame Magazine no. 5
Summer 1980

HUMAN CENTRED SYSTEMS:
AN URGENT PROBLEM FOR SYSTEMS DESIGNERS

First published in *AI&Society* journal, 1987

INTRODUCTION

One of the most precious assets any company, organisation or country possesses is the skill, ingenuity, creativity and enthusiasm of its people. Yet, far from optimising human resources, we seem determined to design systems such as to marginalise human intelligence and tacit knowledge and even seek to preclude them as a form of systems disturbance.

Good design is generally regarded as that which reduces uncertainty. However, if we regard human decision making processes in narrow systems terms we will indeed perceive it to be an uncertainty and are thereby driven to the (false) conclusion that good systems design is that which reduces human intervention to a minimum. A richer way of viewing this would be that the human capacity to handle uncertainty actually contributes to systems robustness.

Conventionally, we do not regard a system as being 'scientific' unless it displays the three predominant characteristics of the 'natural sciences', namely predictability, repeatability and mathematical quantifiability. These characteristics do, however, tend to preclude human judgement, intuition, subjective knowledge, sense of feel and imagination. These problems of systems design are highlighted rather than diminished by many of the new technologies.

One of the founders of cybernetics, Norbert Wiener, once cautioned: 'Although machines are theoretically subject to human criticism, such criticism may be ineffective until long after it is relevant'.[1]

Probably at no time in history has the need for an examination of the underlying assumptions of science and technology been more relevant than it is today. We are, I submit, at a unique historical turning point. Decisions we make in respect of technological developments during the next five or ten years will have profound effects upon the way our society develops; the manner in which human beings relate to machines and to each other; and the relationship between human beings, their built environment and nature itself.

The introduction of computers, their enabling technology and accompanying organisational form has given rise to expressions of fundamental concern as to where these forms of technology are taking us.[2,3] Fears are expressed that by failing to examine the range of technological choices open to us, we are permanently closing off technological options and alternative forms of human organisation which reflect 'a loss of nerve' on the part of engineers and designer.[4] Rosenbrock has described the process by which we eliminate these choices as the 'Lushal Hills effect'.[5]

Integral to these expressions of concern is the notion that we may be about to repeat, in the field of intellectual work, many of the mistakes we made at such enormous cost at earlier historical stages when skilled manual work was subjected to the introduction of high capital equipment.[6] This is perceived to be part of an historical process in which the manner we currently apply science and technology will inevitably lead to deskilling.[7] Against this historical background, techniques such as computer aided design are regarded as a Trojan horse with which to introduce Taylorism into the field of design.[8]

Scientific Management is, as its founder Frederick Winslow Taylor explained, a process in which "the workman is told minutely just what he is to do and how he is to do it and any improvement he makes upon the instructions given to him is fatal to succes".[9]

Initially, Taylorism on the shop floor actually increased the intellectual activity of designers and others in the offices. Taylor himself explained that his technique is aimed at establishing a clear cut and novel division between mental and manual labour throughout workshops. 'It is based upon the precise time and motion study of each worker's job in isolation and relegates the entire mental parts of the task in hand to the managerial staff'.[10]

Seventy years of Scientific Management have seen the fragmentation of skills grind through the spectrum of workshop activity, engulfing even the most creative and satisfying manual tasks such as that of toolmaking. Throughout that period, most industrial laboratories, design offices and administrative centres were the sanctuary of the conceptual, planning and administrative aspects of work. In those areas, one spur to output was a dedication to the task in hand, an interest in it and the satisfaction of dealing with a job from start to finish.

Viewed retrospectively, it now seems naive to have believed that Taylorism would stop at the design office door. It should have been obvious that, the more science and technology ceased to be an amateur gentleman's affair and was integrated into the productive process, the more those involved in it would become part of the workforce itself. Consequently, it has been suggested that as high capital equipment, such as computers, becomes available to them, scientists, technologists and design staff will be paced by the machines and, eventually, their intellectual activities will be subdivided into routine tasks and will be work-studied to set precise times for its synchronisation with the rest of the 'rational process'.

This should have been all the more obvious when it is realised that in fact it was one of the founders of the computer industry, Charles Babbage, who actually anticipated Taylorism, and did so in the field of intellectual work when he wrote: 'We have already mentioned what may perhaps appear paradoxical to some of our readers, namely, that the division of labour can be applied with equal success to mental as well as mechanical operations, and that it ensures in both, the same economy of time'.[11, 12]

The notion of the division of labour and the efficiency which is

The Search for Alternatives

said to flow from it is normally associated with Adam Smith.[13] Adam Smith's specific arguments were anticipated by Henry Martyn almost a century earlier.[14] However, the basic concepts of the division of labour are so intertwined with Western philosophy and scientific methodology that they are identifiable as far back as Plato, when he argued for political institutions of the Republic on the basis of the virtues of specialisation in the economic sphere.

If we are unable to quantify something, we like to pretend that it does not exist. To pretend this, we have to rarify it away from reality and this leads to a dangerous level of abstraction, rather like a microscopic Heisenberg principle. Such techniques may be acceptable in narrow, rarefied mathematical problems but, where much more complex considerations are involved, as in the field of design, they may give rise to very real problems and, indeed, toquestionable results. [As Silver notes:]

> 'The risk that such results may occur is inherent in the scientific method which must abstract common features away from concrete reality in order to achieve clarity and systematisation of thought. However, within the domain of science itself, no adverse results arise because the concepts, ideas and principles are all interrelated in a carefully structured matrix of mutually supporting definitions and interpretations of experimental observation. The trouble starts when the same method is applied to situations where the numbers and complexity of factors is so great that you cannot abstract without doing some damage and without getting an erroneous result.'[15]

PROBLEMS OF HUMAN-MACHINE INTERACTION

Within the design process there is a contradiction at the level of the human-machine interaction itself. The human being may be viewed as the dialectical opposite of the machine in that he or she is slow, inconsistent, unreliable but highly creative. The machine on the other hand may be regarded as fast, consistent, reliable but totally non-creative.[16] Initially these opposite characteristics were perceived as complementary and regarded as providing the basis for a human-machine symbiosis.[17] Such a symbiosis would

Human Centred Systems

however imply dividing the design activity into its creative and non-creative elements. The notion then is that the non-creative elements may be allocated to the machine and the creative elements left to the human beings. This is a Taylorist notion and implies at the level of design the equivalent of separating hand and brain within the field of skilled manual work.

The design activity cannot be separated in this arbitrary way into two disconnected elements, which can then be added and combined like some kind of chemical compound. The process by which these two opposites are united by the designer to produce a new whole is a complex and, as yet, ill-defined and ill-researched area. The sequential basis on which the elements interact is of extreme importance. The nature of that interaction and indeed the ratio of the quantitative to the qualitative depends on the commodity under design consideration. Even where an attempt is made to define the portion that is non-creative, what cannot readily be stated is the stage at which the creative element has to be introduced when a certain part of the non-creative work has been completed. The very subtle process by which designers review the quantitative information they have assembled and then make the qualitative judgement is extremely complex,[18] and much freedom must be left to the designer in doing it.

But there are further problems. The computer can produce quantitative data at an incredible rate. As the designer seeks to keep abreast of this and cope with the qualitative elements, the stress upon him or her can be truly enormous. In certain types of mechanical engineering design examined by the AUEW, instances were found where the decision making rate is forced up by approximately 1900. Clearly, human beings cannot stand this pace of interaction for long. Experiments have shown that the design efficiency of an engineer working at a visual display unit decreases by 30-40% in the first hour and 70-80% in the second hour.[19]

Since employers, particularly those in a non-academic environment, will expect the equipment to be used continuously, the situation can be extremely stressful. Indeed, in 1975, the International Labour Office recommended safeguards against

The Search for Alternatives

nervous fatigue of white collar workers.[20]

An International Federation of Information Processing working party has suggested that mental hazards "caused by inhumanely designed computer systems should be considered a punishable offence just as endangering the bodily safety".[21] Furthermore, in order to optimise the human-machine interaction, measurements are being made of the peak performance age and response time of intellectual workers.[22]

COMPUTER-AIDED DESIGN (CAD)
DESKILLING OR LIBERATING TECHNOLOGY

I have described elsewhere how CAD tends to deskill the designer, subordinate the designer to the machine and give rise to alienation. Indeed, most computerised design environments begin to display those elements which are regarded as constituting industrial alienation, in particular powerlessness, meaninglessness, and loss of self and normality.[23, 24]

Further, computer aided design environments increasingly tend to induce methods of work and decision making techniques which are remarkably at variance with the circumstances and attributes which have appeared to contribute to creativity, both in the arts and in the sciences when viewed historically.[25, 26, 27]

It would be quite mistaken to assume that the routinisation is reserved for the 'hack' part of the design activity, thereby liberating the designer to re-engage in more meaningful work at higher levels of creativity. The situation now is that some of the most creative forms of design activity are being engulfed by this type of 'scientific management'. Elitist designers, steeped in their traditional professionalism, believed (and may still believe) that their creative talents provided an eternal sanctuary against the creeping proletarianisation of all white collar workers. Architects, for example, conceded that there might be problems for aircraft designers and civil engineers, but not for them.

After all, is not architecture the 'queen of the arts rather than the father of technology'?

Human Centred Systems

However, in the relentless drive to control and exploit all who work, the system has not forgotten architects. For them there have been specifically produced software packages such as (appropriately enough) the HARNESS SYSTEM. The concept behind such a system is that the design of a building can be systematised to such an extent that each building is regarded as a communication route. Stored within the computer system are a number of pre-determined architectural elements which can be disposed around the communication route on a visual display unit to produce different building configurations. Only these pre-determined elements may be used and architects are reduced to operating a sophisticated 'Lego' set. Their creativity is limited to choosing how the elements will be disposed, rather than considering, in a panoramic way, the type and forms of the elements which might be used. Thus the creativity of the architect is constrained, and his or her imagination greatly limited; yet it is precisely our ability to use our imagination which distinguishes us from other animal forms.

As Marx observed,

> "A bee puts to shame many an architect in the construction of her cells. But what distinguishes the worst architect from the best of bees is that: the architect raises in his imagination the structure before he erects it in reality. At the beginning of every labour process we get a result that already existed in the consciousness of the labourer at its commencement".[28]

Systems such as HARNESS do reduce architects to bee-like behaviour and do deskill them. It was recently pointed out at a British Computer Society meeting, by architects working for a local authority, that where their colleagues had used systems of this kind for two or three years, they found it extremely difficult to obtain 'normal' architectural design work, since prospective employers regarded them as having been deskilled by using such systems.

There is clearly a need for research projects to establish if these

The Search for Alternatives

problems are inherent in the methodology used in the introduction of CAD, or whether it can be explained away as a mere use/abuse model.

It is obvious that many of the advocates of CAD, particularly those working in academic environments, perceive it as a liberating process and a basis for the democratisation of the decision making process within design.[29] Arthur Llewellyn, former Director of the CAD Centre at Cambridge, repeatedly and so correctly asserted that computers should not be used as a means of diminishing or eliminating designers, but rather as tools for improving their ability to carry out creative tasks.[30]

Unfortunately, this is frequently not so in the harsh world of industrial reality, and the consequences spread far beyond the design area. In some systems, not only will draughtsmen as we now know them be gradually phased out, but some of the most satisfying and skilled work will be eliminated by the resultant numerically controlled (NC) equipment. The optimistic predictions of some industrial sociologists that the worker would be elevated to a sort of manager with the computer providing an overview of the labour process are being denied day by day. In the case of NC equipment, it has been reported that the ideal workers are mentally retarded ones, and a mental age of 12 was mentioned.[31]

HUMAN-CENTRED SYSTEMS – A WAY FORWARD

Clearly it is not sufficient simply to identify these problems. We have to indicate how these systems can be designed otherwise. Two examples may be given to illustrate the possibilities; one in the field of skilled manual work, the other in the field of intellectual work.

Over the past 200 years, turning has been one of the skilled jobs to be found in most engineering workshops. Toolroom turning is one of the most highly skilled jobs of all. The historical tendency, certainly since the war, has been to deskill this function by using NC machines. This is done by part programming – a process by which the desired NC tool motions are converted to finished tapes.

Human Centred Systems

Conventional (symbolic) part programming languages require that a part programme, upon deciding how a part is to be machined, describes the desired tool motions by a series of symbolic commands. These commands are used to define geometric entities, that is points, lines and surfaces which may be given symbolic names.

In practice, the part programming languages require the operator to synthesise the desired tool motion from a restricted available vocabulary of symbolic commands. However, all this is doing is attempting to build into the machine the intelligence that would have been exercised by a skilled worker in going through the labour process.

It is possible, by using computerised equipment in a symbiotic way, to link it to the skills of a human being and define the tool motions without symbolic description. Such a method is called Analogic Part Programming.[32] In this type of part programming, tool motion information is conveyed in analogic form by turning a crank or moving a joystick, or some other hand/eye coordination task, using read-out with precision adequate for the machining process.

Using a dynamic visual display of the entire working area of the machine tool including the workpiece, the fixturing, the cutting tool and its position, the skilled craftsman can directly input the desired tool motions to 'machine' the workpiece in the display. Such a system, which may be described as 'Part Programming by doing', would represent a sharp contrast to the main historical tendency towards Symbolic Part Programming. It would require no knowledge of conventional part programming languages, because the necessity to describe symbolically the desired tool motions would be eliminated. This is achieved by providing a system whereby the information regarding a cut is conveyed in a manner closely resembling the conceptual process of the skilled machinist. Thus it would be necessary to maintain and enhance the skill and ability of a range of people who would work in parallel with the system.

Significant research has been carried out in these fields,[33] yet

The Search for Alternatives

in spite of its obvious advantages it has not been received with any enthusiasm by large corporations or, indeed, funding bodies. That this is so would appear to be an entirely political judgement rather than a technological one.

In the field of intellectual work, Rosenbrock has questioned the underlying assumptions of the manner in which we are developing computer aided systems. He charges, firstly, that the present techniques fail to exploit the opportunity which interactive computing can offer. The computer and the human mind have quite different but complementary abilities. The computer excels in analysis and numerical computation. The human mind excels in pattern recognition, the assessment of complicated situations and the intuitive leap to new solutions. Rosenbrock objects to the 'Automated Manual' type of system since it represents, as he says, "a loss of nerve, a loss of belief in human abilities, and a further unthinking application of the doctrine of the Division of Labour".[34, 35]

As in the case of turning described above, Rosenbrock sees two paths open in respect of design. The first is to accept the skill and knowledge of the designers and attempt to give them improved techniques and improved facilities for exercising their knowledge and skill. Such a system would demand a truly interactive use of computers in a way that allows the different capabilities of the computer and the human mind to be used to the full.

The alternative to this, he suggests, is to 'subdivide and codify the design process, incorporating the knowledge of the existing designers so that it is reduced to a sequence of simple choices'.[36] This, he points out, would lead to a deskilling such that the job can be done by a person with less training and less experience. Rosenbrock has demonstrated the first human enhancing alternative by developing a CAD system with graphic output to develop displays from which the designer can assess stability, speed of response, sensitivity to disturbances and other properties of the system.

If, having looked at the displays, the performance of the system is not satisfactory, the displays will suggest how it may be

Human Centred Systems

improved. In this respect the displays carry on the long tradition of early pencil and paper methods but of course they bring with them much greater computing power. Thus, as with the lathe and the skilled turner, so also with the VDU and the designer, possibilities exist of a symbiotic relationship between the worker and the equipment. In both cases, tacit knowledge and experience is accepted as valid and is enhanced and developed.

In Rosenbrock's case it was necessary to examine the underlying mathematical techniques involved in control systems design.[37] The outcome of his work does demonstrate in embryo that there other alternatives if we are prepared to explore them, and he has suggested that we are at a critical moment when we may close off options which are now open to us.

It will be clear from the foregoing that these alternatives, both in the field of manual and intellectual work, require urgent discussion and research, given the widespread concern at the deskilling and other multiplier effects of conventional 'machine-centred' systems. It is significant that in recent years the major Trade Unions worldwide have shifted from a rather simplistic technological optimism to a growing sense that a re-examination of the underlying assumptions of such technologies is now overdue. The major International Federation of Metal Workers has recently published a report in seven languages discussing, on the one hand, job creation possibilities worldwide and, on the other, linking these possibilities to the use of 'human-centred', skill-enhancing, new forms of manufacturing technology.[38] More recently, several of these concerns have been brought together in a highly significant ESPRIT Project. The objective is to design what will be the world's first human centred computer integrated manufacturing system. The project involves nine partners in three EEC countries. Danish teams will research and ultimately design a CAD system. Partners in Germany will produce the CAP and the British partners, the CAM (with the Greater London Enterprise Board (GLEB) acting as prime contractor and coordinator of the project). It was accepted from the outset that science and technology are not neutral but rather that they embody assumptions

The Search for Alternatives

of the society which has given rise to them. Furthermore, that it is mistaken to assume that there is only one form of science and technology. In the case of manufacturing technology, this might be regarded as 'American Technology'. One should regard science and technology as aspects of culture, and just as culture produces different music, different literature and different language, why should there not be different forms of technology? In this case the object is to design a manufacturing technology which will reflect European cultural, social and political aspirations. Many of the European expectations tend to exist more in the rhetoric than in the reality. Even at a rhetorical level, the idea of designing a system which embodies concerns such as motivation, the dignity of the individual, the enhancement of skill, a respect for quality and support for those organisational forms which diffuse decision making as widely as possible. The project[39] has now been supported by the EEC to a level of £3.8M over three years. One of the objectives throughout will be to ensure that the qualitative or subjective judgements of the worker will be treated as valid scientific knowledge and the worker will dominate the machine rather than the other way round. The systems design will be such as to build upon and enhance the skill, ingenuity and tacit knowledge of the worker. It will reflect the European reality of highly skilled workers and small to medium sized companies.

It is held that such a 'human-centred' approach could be applied more widely to system design work undertaken by the Greater London Enterprise Board in a number of projects that are likewise based on a human centred approach.[40] Work is proceeding on an Expert Medical system in which the object is to diffuse knowledge outwards into general practice and the community. The database will be so structured as to render visible the assumptions to the general practitioner rather than to concentrate knowledge in the hands of a small élite of consultants. Thus it is hoped it will be possible to democratise decision making, sharing it between the general practitioner and the consultant. More particularly the presentation of the data in the surgery is such as to provide both the patient and the medical practitioner with a range of different treatment options. This, firstly, avoids the major defect in Western scientific methodology; 'the notion of the one best way'. Also it enforces a dialogue between the medical practitioner and the patient, thus providing a framework in which this relationship can be more democratised. Likewise, an interactive video disc project at GLEB seeks to build upon the knowledge and culture of ethnic minorities, rather than providing them with a series of 'commands'. It is hoped that such work on Expert Systems will be ultimately integrated into manufacturing systems since, of course in the best sense, a skilled worker on the shop floor is as much an expert as is a medical practitioner or a lawyer. In designing such systems no attempt is made to reduce the knowledge of the practitioner to a totally rule-based system. It is accepted that there are the facts of the domain and that these provide the basis for a rule based 'core', but that surrounding this there is tacit knowledge, heuristics, fuzzy reasoning, intuition, imagination, and that these cannot or should not be reduced to a rule base. Thus the systems are designed to create a dialectical interaction between the rule-based core and the rest. In designing such systems an attempt is made to concentrate on that part of the cybernetic loop which is knowledge and wisdom based rather than on the data and information end. Thus we are concerned with minimising noise in the system and enhancing signal.

Conclusions

This work is as yet at an early stage. It is however hoped that projects of this kind will begin to bring about paradigmatic shifts in our concepts of how systems can be designed in future, and that they will provide an environment in which such systems can be carried out in practice. It is our view that the problems and contradictions of the existing systems are now so great that the search for alternatives is a matter of considerable urgency and that 'human-centred' systems provide one interesting starting point.

The Search for Alternatives

Notes

1. Weiner, N. (1960), *Science* 131, 1355.

2. Weizenbaum, J. (1972), 'On the impact of the computer in society - how does one insult a machine?', *Science* 176, 609-614.

3. Weizenbaum, J. (1976), *Computer power and human reasoning,* W.H. Freeman, San Francisco.

4. Rosenbrock, H. (1977), 'The future of control', *Proceedings,* 6th IFAC Congress, Boston, 1976, *Automatic* 13, 389-392.

5. Rosenbrock, H. (1979), 'The redirection of technology', *Proceedings,* IFAC Conference.

6. Cooley, M.J.E. (1972), *Computer aided design - its nature and implications,* AUEW, London.

7. Braverman, H. (1974), *Labor and monopoly capital - the degradation of work in the 20th century,* Monthly Review Press, New York.

8. Cooley, M.J.E. (1976), 'CAD - a trade union viewpoint', *Proceedingf,* CAD '76,308-312. IPC Press, Guildford.

9. Taylor, F.W. (1906), *On the art of cutting metals,* 3rd edn revised, ASME, New York.

10. Taylor, F.W. (1903), *Shop management,* New York.

11. Babbage, C. (1906), On *the economy of machinery and manufactures,* Kelly, New York.

12. Urquhart, A. (1855) 'Familiar words', in K. Marx, (1961), *Capital,* 1, 363, Lawrence & Wishart, London.

13. Smith, A. (1937), *The wealth of nations,* Random House, New York.

14. Martyn, H. (1701), *Consideration upon East India Trade,* London.

15. Silver, R.S. (1975), 'The misuse of science', *New Scientist,* 166, 956,555.

16. Cooley, M.J.E. (1987), *Architect or bee?,* Chatto & Windus, London.

17. Licklider, J.C.R. (1960), 'Man computer symbiosis', *IRE Trans Electpon, 2,* 4-11.

18. Archer, L.B. (1973), *Computer design theory and handling the*

qualitative, Royal College of Art.

19. Bernholz, A. (1973), *Proceedings,* IFIP CAD Conference, Eindhoven.

20. ILO Convention No 120, xvi, 60.

21. 'Human choice and computers' (1974), Proceedings, IFIP Conference, Vienna, Report HCC, 5.

22. Cooley, M.J.E. (1979), Computers and politics: state of the art report, Infotech, Maidenhead.

23. Seeman, M. (1959), 'On the meaning of alienation', American Sociological Review, 24, 6, 783-791.

24. Schacht, R. (1970), Alienation, Anchor Books, New York.

25. Beveridge, W.I.B. (1961), The art of scientifc investigation, Mercury Books, London.

26. Eiseley, L. (1962), The mind as nature, Harper & Row, New York.

27. Fabin, D. 'You and creativity', Kaiser Aluminum News, 25, No3.

28. Marx, K. (1974), Capital, 1,174, Lawrence & Wishart, London.

29. Maver, T. (1976), 'Democracy in design decision making', Proceedings, CAD '76, 313-318, IPC Press, Guildford.

30. Llewellyn, A. (1973), Computer aided draughting systems, 4,94, IPC Press, Guildford.

31. Reported in 'The American Machinist', July 1979.

32. Gossard, D. & von Turkovitch, B. (1976), 'Analogic part programming with interactive graphics', Annals of the CIRP, 27, January.

33. Gossard, D. (1975), 'Analogic part programming with interactive graphics' PhD thesis, MIT, Boston.

34. Rosenbrock, H.H. 'The future of control', op. cit.

35. Rosenbrock, H.H. (1977), 'Interactive computing: a new opportunity', Control Systems Centre Report No 388, UMIST, Boston.

36. Rosenbrock, H.H. (1974), Computer aided control system design, Academic Press, London.

37. Cooley, M.J.E. (1980), Architect or bee? - the human/technology relationship, LangleyTechnical Services, 95 Sussex Place, Slough SL1 INN, UK.

38. Cooley, M.J.E. (1984), Technology, trade union and human needs, International Metalworkers Federation, Geneva.

39. Cooley, M.J.E. & Crampton, S. (1986), 'Criteria for human centre CIM', Proceedings, University of Bremen.

40. Various douments and papers from Technet Ltd, GLEB, 63-67 Newington Causeway, London, SEI.

First published:
AI & SOCIETY, VOL. 1, 37-46 (1987)

TECHNOLOGY AND INTERNATIONAL DEVELOPMENT

First published by Univeristy College Cork, 1992

TECHNOLOGY AND DEVELOPMENT

The gap between the potential that technology provides for society and its reality is developing into a chasm. We have, on the one hand, control systems which can guide a missile to another continent with an accuracy of a few centimetres, yet the blind and the disabled still stagger around our cities in very much the same way as they did in medieval times.

When we get a spin-off from that technology we end up with something as elegant and complex as Concord, yet the very same society tolerates a situation in which every year hundreds of pensioners die because they cannot get a simple effective heating system. We have people optimising car bodies so that they are aerodynamically stable at 120mph, but the average speed of a car to the centre of New York at peak travel times is 6.2mph. Apparently it was 8mph at the beginning of the century when they were horse-drawn. The banks and multi-national corporations are working on international networks so that we can send messages round the world in fractions of a second, but it now takes longer to send an ordinary letter from Washington to New York than it did in the days of the stagecoach.

Another contradiction is the tragic wastage our society is now making of one of its most precious assets. Much is said, correctly, about the wastage of energy and materials, but precious little is

The Search for Alternatives

said about the wastage of skill, ingenuity, creativity, energy and enthusiasm of ordinary people. I must say I have never succeeded in meeting an ordinary person. All the people I meet are quite extraordinary with all kinds of skills, dreams, hobbies, fears, and ambitions. Very seldom do mechanisms exist for them to give vent to their individual ideas and concerns. There is a growing hostility to science and technology as it is presently practised in our technologically advanced society. As an engineer and a scientist, I think it is sad that some of our most able, creative, constructive and sensitive young people will no longer study science and technology. They see it as repressive or running counter to the very best in the human essence.

Against that background I would like to look, constructively, I hope, at a number of attempts to deal with these problems and to stimulate economic development. I would like, before I do that, to look briefly at the context in which I will be talking about this.

Reconstruction

The context is one of massive and disorienting reconstruction of industrial society. That reconstruction displays three predominant historical tendencies. I would suggest that these tendencies have been discernible in Europe over the past 400 years and I do not accept that what is now bearing down on us is a technological bolt from the blue. It seems to me that we have been asking the wrong questions of science and technology and I find it therefore not surprising that we get the wrong answers.

The three predominant historical tendencies are, first, changing the organic composition of capital and making everything capital and energy intensive rather than labour intensive. That, as I will show, is giving rise to fairly massive growing structural unemployment. The second is a shift from the analogical to the digital, the way in which we perceive our world and relate to it is changing dramatically. Thirdly, it is a process in which human beings are rendered passive and the systems become more active. You will probably recall that someone once said, "The more you give to a machine the less there is left of yourself". That is as true of areas of new technology as it is true of people giving of themselves to the nation state.

I would like to briefly give a sense of this rate of change and I will then talk about some attempts being made to deal with them. First we are experiencing a massive transformation of industrial society. If we look at the United States over the last few hundred years the transformation has been extraordinary. Agriculture has increasingly mechanized and increased its use of chemicals. There are calculations to show that the chemical input of the food is actually greater than that of the energy output. This is something we are going to have to address in the long term. The level of chemical input is so great that in France it was recently pointed out that milk from a mother's breast is so polluted it would be illegal to sell it in a bottle, yet, in every language, in the vernacular, we have terms like "As pure as a mother's milk". In two hundred years we are just beginning to see the multiplier effects of these transformations.

Those being displaced from the land were going into manufacturing industry where the increases in numbers of people continued up to the 1970s. In the meantime the white collar administrative area has been growing, but that in itself is being transformed by increasing use of word processing systems, computer design systems and so on. These changes are laudable, in some circumstance, but deeply problematic if there isn't a structure in society to deal with them. The transformations are enormous in the fields of electromechanical industries. A telecommunications plant which used to require 26 workers, requires 10 workers with first generation electronics and with systems X it will be 1. A productivity increase of 10 to 1. The cost of robotic equipment which is replacing human beings is dramatically reducing. Ten years ago it would have required the wages of 10 workers, five years ago 5, now about 3 and in around three years' time, the cost will be of 1 worker. The cost is going down and the incentive to replace people by machines will grow quite rapidly.

The Search for Alternatives

UNEMPLOYMENT

In all of the EEC countries we now see the basis of unemployment going up. There is a prediction from the EEC that if we could maintain our present growth rate, which I think is highly questionable, we would still have something like 20 million people out of work by 1992. The rate at which this is happening is very difficult for us to conceptualise, especially when we are in the middle of it.

There are frantic attempts being made to produce computers which will replace the human mind. It would have required a volume as great as the Greater London area in 1950, by 1960 it was down to the size of the Albert Hall, in 1970 a London bus, in 1980 a taxi, 1990 a TV set and by the year 2000 they believe they will be there. Of course, those who say that, don't understand the difference between a rule based system and the human mind with its unrealistic fuzzy reasoning, all those things which make us unique and precious and distinguishes us from machine based systems.

So we see the basis of unemployment tending to go up but the jobs vacant remaining fairly static. The gap between these two will get bigger and bigger and will cause all sorts of structural problems. In West Germany they are attempting to assert the right of all to work. In the absence of that, even in the tranquil industrial relations of West Germany, there has been a whole series of disputes about a shorter working week, longer holidays and more leisure time. But, even if we were to get that, it seems to me that would be dealing with a minor part of the problem. We need something much more fundamental and structural before we can address a problem of this scale and magnitude, otherwise the riots we have seen in inner-city areas, and which we saw in the United States in the 1960's, will be repeated.

Technology and International Development

GREATER LONDON ENTERPRISE BOARD

In Greater London, there were at the time about 350,000 people out of work, and an attempt was made to look at models to deal with these growing problems. One of the models used was the Lucas workers' plan for socially useful production where, faced with the growing structural unemployment, the workforce took the initiative and did not passively wait there whining about a problem. One of the things which was done was to set up the Greater London Enterprise Board (GLEB). It was provided with £32 million a year, which might seem a lot but actually that is about £3 per citizen per annum, which is petty cash compared with the money used by the multinational corporations. With that money, in three years, it was possible to create 4,000 jobs directly and many more indirectly. Those jobs cost about £4,700 each, whereas, if someone is out of work in Britain, and has got a couple of dependants, it would cost about £7,000 per annum to keep them, and that figure repeats itself each year.

Now I freely admit that 4,000 jobs is quite small but what is important about it is that it shows in embryo what is possible if these resources are made available. One of the mechanisms for dealing with this was to look for new products, new companies, new activities, and new services which might allow people to become engaged in productive and, hopefully, more long-term profitable activities. It was felt from the onset that this could not be done from on high, and there were adverts in the papers, one of which said that the key of the future was the imagination, knowledge and creative energy of the people of London.

Discussions took place with the seven polytechnics and three universities throughout London as a result of which technology networks were set up. Some of these covered geographical areas so that there was a North and East network, there was a South-East network, and there has recently been established a West London network. There were also product networks, an energy network, a transport network, and a new technology network.

The Search for Alternatives

THE TECHNOLOGY NETWORK

The technology network building was located close to a university but not on a university campus. We find that a university is very alienating for the unemployed, for practical people, for the women's movement, and for those whose experience of life is real and practical. Actually I happen to believe it is also alienating for the students, but they don't really have any alternatives. It was also important to ensure that there were four times as much workshop space as office space in each of the centres. In my experience, activity that touches local or central government can degenerate overnight into report writing. There are those who believe that a report constitutes work done, whereas to me, as an engineer, it's merely a guide to something real that is subsequently going to happen.

It was also important to try and create an atmosphere in which practical, ordinary people felt at ease. I think we do enormously confuse linguistic ability with intelligence. We are over-impressed with what people say and what they write. Most people express their intelligence by what they do and how they do it, rather than the way they write and talk about it. I have been ever amazed by the way an academic can make a doctoral thesis with about 970 references in describing in unusually boring detail what a skilled worker will do day in and day out without even thinking about it. It was important to create an atmosphere in which people felt at ease so that they could give vent to their creativity in a whole range of practical ways.

BILL

The kind of people who came in were like Bill who had been a skilled sheet metal worker at Ferrari's. He could take a flat sheet of steel and, with hand, eye and brain co-ordination, could literally make up the wing of a crashed Ferrari car. He said he had an idea for the alternative to the Sinclair C5. You may remember it was a small bicycle which had a very good basic idea. This was that if it

Technology and International Development

went at less than 15 miles per hour, and had a power pack less than 200 watts, it would rate as a bicycle/tricycle rather than a motorcycle, so you didn't have MOT testing, drivers' licence and so on. Bill came in and said he had an alternative which was a power-assisted bicycle.

The first thing he did was quickly build up on a breadboard a small circuit. He really would have difficulty describing the theory of that but he was certainly able to get it fairly close. He then went to Imperial College and asked them to facilitate it through the network, for them to fine tune it for him, so the initiative was with him rather than with Imperial College. The academics were supporting and optimising something which he had initiated, rather than the other way round.

He also worked on a special motor, a permanent magnet motor, that has now been developed and will give an energy saving of about 13% over the best motors which are available worldwide. That again was done in a very practical fashion. Sadly not one British company seems to have the energy and willingness to take up this product nor does there seem to be the possibility of setting up a large-scale co-operative to deal with it. So it looks as though it will go to a German company.

In building this he didn't massively calculate everything, he quickly built a series of prototypes. He didn't even accept that he should necessarily have a chain drive, so produced a small drive wheel with a V belt. He developed a special suspension mechanism at the back which pivots around a point and compresses a spring. In three months he had his second and final prototype and in the fourth month he had a product which would give you 922 miles per gallon equivalent. It has a small microprocessor so that you have an initial velocity of 3 miles per hour before you put on the power supply and it is capable of an incline of 1 in 6 plus. It has an aerobatic battery so that you can turn it upside down to repair it, and you can use it as a bicycle and get it to assist you when you want to go up an incline.

The possibilities are enormous. One can imagine a cyclist being at last taken seriously in cities, as they already are in Amsterdam

The Search for Alternatives

and elsewhere, so it would provide an environmentally desirable form of transport. It would mean that older people who had given up cycling would be able to use power-assisted facilities when they meet a steep incline. More people can take up cycling again. There is this obsession with health and exercise at the moment. Instead of sitting on one of these ridiculous exercise bikes in the comfort of your own front room, where you pretend you are cycling although you are going no place, you could actually use one of these to travel around the city areas.

When this was shown on *Tomorrow's World*, the Italian company, Testy, came over two days later with its Managing Director wanting to manufacture them and pointing out that there was a market of a quarter of a million a year in the People's Republic of China. They want to get away from the little Mopeds they have and believe that battery power as an interim technology would be more desirable.

It looks as if, because of the lack of enterprise and drive, products like this although designed and developed in Britain, will be going abroad. There is a lack of that kind of "get up and go", the enterprise characteristic, and I don't mean that in a narrow individual sense, I mean it in a much wider social sense. One can imagine how a community could escape large-scale unemployment making these and it would be environmentally and otherwise desirable.

THE POSSIBILITIES

In the whole area of energy enormous possibilities exist. In council estates, in community groups and elsewhere one could develop various forms of heat recovery systems and heat pumps using natural gas in internal combustion engines. If you do this you get something like 2.8 times as much use of energy than you would get if you burnt the gas directly and, if these were linked to creative rehab schemes in inner city areas, one can see the enormous possibilities that begin to emerge. Although early models of these were developed in Britain there is now a German company exporting them into Britain.

We are importing when a lot of the original work on these kinds of development has already been done in Britain. A number of wind generators were developed and even the humble steam engine, re-examined in light of new materials has produced a unit which is now being tested on a small boat on the Thames. The possibilities in Third World countries would present a very significant market as well. We are certainly going to have to think of the problems of energy and energy consumption in the long term.

Transport

Our colleagues in the Midlands set up a co-operative to make sections of small vehicles which were designed in such a way that you really could repair them. It uses a donor mini-engine and you could build one of those in about 180 hours. It caused enormous excitement with young people working alongside a skilled mechanic because there is a very fast feedback. Within a month or so they are driving around in a vehicle which they have helped to build.

The basic idea, it seems to me, is really quite important. We are certainly going to have to look in the long term at the notion that we can go on producing throw-away cars. If you throw away a car under ten years or 80,000 miles or whichever comes first, you are broadly throwing away the amount of energy required to drive it 80,000 miles. We need to think creatively of alternatives.

Communities, for example, which have a tremendous inheritance from the Victorian engineers, a guided transport system, should be looking at creative ways of using it rather than allowing that to be vandalised, and I mean state vandalism rather than local young people. One of the possibilities is to have road/rail vehicles of this kind so that you have a truly integrated system. The conversion of old buses could be carried out in fairly simple workshops given that the complex components could be manufactured elsewhere.

The Search for Alternatives

What I am talking about is lateral thinking, not just going in a straight line. Someone asked what would happen when one of the tyres on a vehicle burst when running on the track at sixty miles per hour, a very good practical question. Now the bureaucratic answer would be to put more wheels on, but one of their colleagues asked whether it would be possible to design a tyre which would have the characteristics of a pneumatic tyre, but which cannot burst or deflate. This would dramatically increase safer possibilities on motorways where many of the accidents are of cars destabilising with one wheel going off, aircraft landing with a wheel bursting and so on. Yet no British company has been willing to take this up – it is a German company which is now making them in short batch production.

So once again these good ideas are coming forward. A tremendous amount could be done in the field of design in the medical field of disability. While I was at British Aerospace some of my colleagues heard that children with Spina Bifida were literally crawling around on the ground. They designed a little hob cart which is of therapeutic value to support the spinal column at the point of weakness as the child propels itself along. An incredibly simple little product which completely transformed the life of a disabled child and the designer, Mike Parry Evans, a leading systems designer in the aerospace industry, described to me the enormous personal pleasure when he brought that down and saw the happiness on that child's face.

SOCIAL MATTERS

All of these kind of things will begin to open up what I call social markets. We should not just think in terms of conventional markets and the demand for grade A or B, rather we should be going to groups demonstrating new products and winning support for the production of those. Whole new markets can be opened up in these areas creating all kinds of interesting and exciting jobs for communities and co-operatives.

Now when you are talking about this sort of thing, politicians and others will ask, "How can we afford to pay for it". I would ask, how can we afford not to pay for it? If you put somebody out of work at the moment the Department of Social Security payment is 60% of the average industrial wage and the loss of revenue to the nation state is 40 % . Together that is 100% of the average industrial wage and that is just the first two economic multipliers.

If you take into account the social multipliers of drug taking, interpersonal violence, the decline of inner city areas, the illnesses which are directly related to unemployment, we get some measure of the cost of unemployment at the level of the nation state. We should be campaigning to have the resources which are used to degrade people in the dole queue transferred into activities which, even if they only break even, would in my view be justified. Society at least is ending up with a product which is needed for re-wealth rather than the degradation and suffering of unemployment.

HEALTH

One can develop, by using expert systems, all kinds of new services to the community. We have worked with St. Thomas's Hospital with knowledge from the consultants which can be sent outwards, back into general practice and the community, in such a way that the decisions of the consultants are made visible to the general practitioner. This means that patients in an acute state of illness do not have to go into those factory-like alienating hospitals but can be treated in the local community. We have deliberately built in a data base management system so that the medical practitioner and the patient are confronted with different treatment options. There is an enforced dialogue between the medical practitioner and the patient so you can actually build in democratising routines.

Another possibility in areas like this is dealing with the major killers in western society, those illnesses associated with the heart and blood systems, cardio-vascular illnesses. We did some work with St. Thomas's Hospital and the East Dulwich Hospital where

The Search for Alternatives

they developed a technique which could scan the body periphery in 20 minutes. This avoids the patient having to spend a week in a National Health Service hospital with a dangerous and painful diagnostic procedure which includes an intrusive surgical technique. The prototype was built in the lower section of a London bus. You can imagine how people out of work in the coach-building industry could begin to make dedicated vehicles for this kind of thing.

This could have significance in terms of dealing with health problems on the same scale as X-rays did in dealing with tuberculosis after the war. There is the possiblilty of going to areas of particular needs to people who are vulnerable to these types of illnesses. We could have a National Health Service rather than a National Illness Service. We have a National Service you go to when you are ill and not when you are healthy. If it tried to ensure that people stayed healthy the possibilities would be truly enormous.

New Forms of Employment

It may seem that the thing I have been talking about is some sort of very radical proposal. In fact, in West Germany the great Metal Workers' Union, which numbers some 2.8 million people, together with two of the major local governments, are now considering ways of creating new forms of employment and developing new products by similar types of proposals. They say it should be based on a rational use of work, information, energy and materials, and that it should be done in conjunction with people leading to a shorter working week. The initiative is for a total programme for West Germany of 50 billion DM to develop new products and new ways of creating employment. Some of the proposals are very close to the kind of things I have outlined. They question whether it is going to be possible to go on producing throw-away cars. Are we going to have to tolerate the levels of pollution? Are there different types of vehicles we could be developing? They are looking into key areas of transport systems which are run both on road and rail, fairly close to the sort of thing I was talking about, as part of a very definite job creation scheme.

They are also looking at other ways of linking energy and materials and for the shipbuilding industry which is in serious crisis. Instead of trying to compete with the Japanese they have been looking at the possibility of very large vertical wind turbines as power units for their ships. They are even considering, in the long term, what one should do about the North Sea Oil platforms. It is extraordinary that no one is thinking about what is going to happen to them. When they run out of gas and oil within the next few years it would be possible perhaps to have large-scale wind generators and take the supply inland from the platform. They are also looking into the possibility of developing air ships. This was proposed in the Lucas plan but at a much more mundane level. They are questioning the whole way in which waste will be disposed of in cities and they are looking at the tremendous energy recovery systems which are now using fairly straightforward technology. These are the kind of options which every community group in my view should be discussing and these are beginning to be discussed in West Germany.

SUNRISE, SUNSET

Now in discussing how we are dealing with these growing structural problems it is a mistake to believe that one should expect that there is on the one hand sunset industries, sunset communities and sunset activities, and on the other hand sunrise equivalents. It is frequently the case that it is possible to transform a sunset industry by introducing aspects of new technology. In West London where we have the old electromechanical skills, one group is developing a wide body jet loader. By linking them up with an electronics company, with people who knew what they were doing, it was possible for them to see electronics as controlling something which they understood very basically as a mechanical system. In other words if we demystify these kinds of things we will find that a lot of skills which were perceived to be irrelevant or lost can be built upon.

There is a programme in Essen, West Germany, where they have very interesting educational environments where people learn by doing. The whole emphasis is on education. My own hierarchy of learning would be that you programme a robot, you train a dog, and possibly a soldier, and you provide educational development environments for human beings. They deliberately develop interactive systems so that people can learn by doing. It is incredible that people who have had difficulty with electronics in a conventional sense, can deal with it quite easily because hardware and software has been made available for them to do so. Now there is grave concern about the de-skilling of some of the new industries. There are now systems being introduced where the drawing as a means of communication between design and manufacture is disappearing. There will be a real discontinuity. For thousands of years, since we drew on cave walls, we have taken two-dimensional data and we have been able to conceptualise what that would be like as a three-dimensional artefact. By downloading straight onto the machines they have recently suggested in the United States that the ideal workers to operate the so-called advanced technology are mentally retarded. They recommend a mental age of twelve years. Now, had the objective been to create work for the mentally retarded, that clearly would have been laudable, but what we are witnessing is the destruction of seedbeds from which the future generations of skill and knowledge are to come.

Culture

Now, rather than complain about that, we have to have an enterprising and creative response. To look at what happens when people work on machines and see if it is not possible to restructure them in a way that builds on human skill rather than marginalising it. We went to the EEC and said quite unashamedly we did not regard science and technology as neutral. We regarded it as a part of culture and just as cultures produced different language, different music and literature, why shouldn't there be different forms of technology which reflect the so-called European aspirations. Aspirations such as creativity, initiative, the sense of quality, the sense of dignity and so on.

Together with colleagues in Germany and Denmark we have begun to build the system which will respond to a skill of a skilled worker. Now most of us completely underestimate the vast band of knowledge we bring to bear in anything we do. People say, for example, "It is as easy as crossing the road". As an Aerospace Technologist, I have been ever impressed on people's ability to cross a road. They go to the edge of a road, they look up and down in both directions, they work out the rate of change of image of the oncoming vehicle, they call up a massive memory bank and note if it is a Mini or a Bus, hence the significance of the rate of change. From the rate of change they compute the velocity of each one separately, then they work out the closing velocity of the two. Meantime they are looking at the width of the road, they are working out their own acceleration, the vector velocity of the system and then they cross the road. And that is an example of something being extremely simple.

If you see a nurse at work or a skilled fitter then you see vast bands of knowledge at work, so we have been developing systems in which you build on that skill rather than marginalise it. If one takes a most simple set routine, whether that workpiece would stay in the chuck or not, you can work out a bending moment around the jaws. If the bending moment here is greater, then it will stay in the chuck. What the skilled worker does is take a chuck key, put it in, feel how tight it is and make a qualitative judgement, which in our experience has always been correct. So we have taken that as a basic assumption, built it into the system and we have designed interfaces in which the human being handles the qualitative subjective judgements and the machine the quantitative elements. You end up with systems which are more labour intensified than they would have been but which build on existing skills and enhance and extend these skills rather than marginalise and displace them.

The Search for Alternatives

HUMAN QUALITIES

That work started two years ago and since it has been going Rolls Royce of Leavesden and BICC have joined it. Employers are beginning to realise that one should not just think about production but the reproduction of knowledge. Where are the next generations of skill and knowledge going to come from if we allocate all that activity to machines? If we deny human beings the capacity to cope and handle uncertainty?

One of the deep problems of our present system's design is we only regard it as being scientifically designed if it displays the three predominant characteristics of Western science and technology, which are predictability, repeatability and mathematical quantifiability. That by definition precludes intuition, subjective judgement, tacit knowledge, imagination, and all those things which make it precious to be human beings.

It is significant that the competition in Japan is moving away from these huge workerless factories. They are realising that if you have a highly synchronised co-ordinated factory and something goes wrong the whole thing turns into de-synchronisation and you get massive confusion. If you have skilled people there who can understand what is going on then immediately something goes wrong you can begin to correct and modify.

TRAINING

There has recently been a report comparing down time in Britain and West Germany. It was found that the down time was massively greater in British factories than in German ones. The first assumption was that they are dealing with clapped out equipment and that is the reason. When we compared like with like it was shown that down time was still an order of magnitude greater.

In West Germany they have highly skilled people who have been through an apprenticeship with all kinds of additional courses and training. The system was something they were capable of tuning and correcting. They could anticipate systems failure. Sadly many in Britain are the product, in my view, of Mickey Mouse training schemes. Narrow schemes which give you specific skills for one narrow little machine. Given the rate of technological change you are left with nothing in two or three years' time when that has gone. An incredibly short-sighted view of skill and skill acquisition. So big employers are now realising the advantage of having people who are really skilled and know what they are doing, and we are beginning to develop machines that will enable them to do that.

PRACTICAL KNOWLEDGE

The great Duomo in Florence, built by Brunelleschi, has actually got 21,000 tons of material and they designed special machines to lift it. They worked out the logistics of getting the material on the site. Nobody came to that site in a white collar and said – "I'm the Manager, I will tell you how to do it". The logistics and the management function was integral to that labour process. The form of the Duomo is one of the most complex structures in Europe, an incredible tribute to human ingenuity, but already at that stage they were beginning to separate the intellectual part of the work from the manual. About that time, for the first time in all the European languages, the word design appeared.

It is possible to develop forms of technology that link back to a more creative way of doing things and yet we deny the significance of that kind of intelligence and knowledge. Even Leonardo da Vinci was deeply upset about the way he was being put down by the so called theorists. He said on one occasion, "They would say that not having learning I do not properly speak of that which I wish to elucidate, but do they not know that my subjects are the better illustrated from experience than by yet more words".

We are denying the significance of practical experiential knowledge. If we are going to solve these huge structural problems now growing in society, we are going to have to release the energy of masses of people. An apprenticeship in medieval times, belonging to a craft guild, was not just a transmission of narrow technical skills, it was the transmission of a culture. A sense of feeling, intentionality and confidence, which is what many of our schemes lack. We do not convey this sense of confidence because young people do not have the opportunity to work alongside people who do have those capabilities so that they are transmitted.

THE INTERNATIONAL CONTEXT

We must view these problems, wherever we are, in an international context. First, in a context of 1992 in the EEC. Many people are suggesting that this will automatically mean that everything is going to be better. There are grounds for believing, indeed many of the bureaucrats in Brussels already believe, that there could be an expansion in the polarization we have seen in individual nation states.

In Britain there are centres of commercial activity in Manchester, Birmingham, Newcastle and London. People are driven from communities in Wales, Scotland and elsewhere to work in these centres. The same thing could be repeated at the macro level in the EEC with parts of Germany concentrating on the machine tool industry; France on telecommunications and London on finance capital, with the rest of the periphery dramatically suffering. By the periphery they mean the whole of the North of England, Ireland, Denmark, Northern Germany, parts of Greece, and Southern Italy. So it is very important that we start to plan the sort of structures we want in our communities otherwise the agenda is going to be set by others. More particularly, we have to begin to think in a more long-term way. Now I don't believe in arid stage planning. What I am talking about is dynamic involvement of people in beginning to define what kind of future they want.

In Japan they have already started discussions for the 21st century and forming 21st century associations in each area. In each of these associations you have local government, the universities and schools, local industry, private citizens and public organisations, like Trade Unions and Chambers of Commerce, all beginning to discuss what the 21st century could and should be like. Now I have a lot of problems with Japanese society. It is not as we are given to understand from the press and I could bore your with details of the deep problems in Japanese society. What I am referring to here is an attempt to look at the future.

Osaka

In the city of Osaka they have set out objectives to create a city of beauty and a sunny environment, to provide the citizens with a full range of cultural activities, to encourage future industry, and to make the city an international centre. They are keen to link the development with an historical past, so there isn't just a massive discontinuity, and in particular they seem to be beginning to address the issues of the environment. Pollution in Japan is quite unbelievable in certain areas.

They are talking about looking at new areas, they are looking at breathing new life into reclaimed parts of the city. One can imagine breathing new life into the inner-cities in Britain. I don't think the mechanisms are developed yet to do it but it should at least be an objective.

In Osaka they are talking about doing it by using high-level technology of various kinds and they talk about promoting exchanges between people, goods, technology and culture. Now I think the reality will be quite different, having witnessed Japanese society close up for a period of time, but the fact is that such a discussion has been initiated there. I find very little of that discussion in our universities or schools whereas they are trying to involve their children and communities in sharing what they want.

Third World Context

One of the most appalling features of the world today is the awesome gap between what science and technology could provide

The Search for Alternatives

for society – its potential – and what it actually provides. The pinnacles of our technological endeavour allow us to develop control systems which can guide a missile with pinpoint accuracy to a remote target, yet the blind and the lame stagger around in much the same way as they did in medieval times. With extraordinary flight-path synchronisation, we can refuel aircraft in flight, transferring thousands of gallons of fuel from one to the other. Yet in Third World countries, vast numbers of our fellow human beings die terribly painfully – simply from dehydration. Many scientists and engineers discuss (usually well out of earshot of their paymasters) the disgraceful funding priorities that result in such a distorted use of their skills and abilities.

But for the majority, there is the one dimensional fascination and challenge of high technology, the conquest of space, the manipulation of data, the control of nature and the exploitation of natural resources including oil.

Against these self-defining Olympian tasks, providing food, shelter and warmth for the less fortunate sectors of humanity is made to look incredibly boring and downright pedestrian. More sinister still is that moral bolthole, the occupational excuse that as scientist we are only doing what we are told. In the frantic linear drive forward to research, design and develop ever more complex systems, any pausing to question who it is that sets these priorities and why, appears at once to be disloyal, diversionary and even subversive.

The Gulf War and its aftermath have now propelled these huge issues into the front rooms of millions of ordinary citizens worldwide. Night in, night out, we sat mesmerized at the awesome precision of delivery systems which gave a new and perverse meaning to the term "surgical". There was also the movement of hundreds of thousands of troops from the US and elsewhere to the Middle East; vast flotillas of ships brought into position; advanced communication networks sending messages from the field commanders to Washington and London; logistical support of mind-boggling complexity. What a spectacular contrast this is to the relatively puny efforts to get aid to the Kurds. From the onset of their plight to April 12th, the British had flown 64 missions to drop supplies to the Kurds. In the same length of time during the war, they had flown 28,000 bombing missions. For the Kurds, when aid eventually did arrive, there was the chaotic plummeting descent of three-tonne pallets from Hercules C130s, sometimes killing those below.

The technology of the military/industrial complex is essentially a "seek-and-destroy" one, not a "succour and sustain" one. Surveillance technologies can intercept the faint signals emanating from enemy control centres, but we do not hear the dying whimpers of a dehydrated child. In the theatre of war, we can move elaborate field hospitals and vast repair workshops into position, but we seem unable to provide basic shelter for those dying of hypothermia each night on the exposed hills on the Iraqi border.

Some relief agencies suggest that a mere one per cent of current spending on weapon systems would provide the basis for dealing with these problems. Frequently, the technology required is rudimentary: irrigation equipment with simple pumping and filtering devices could transform the lives of many. In some cases these could be wind-powered pumps with simple conduits, using local materials. Appropriate, locally repairable agricultural implements could lay the basis for a reasonably predictable supply of food. The objective throughout would be to support these people in gaining their economic independence. Our approach to such collaboration with developing countries should be very modest indeed, bearing in mind the mess we have made of our own science and technology in the military sense, our more conventional technology is often nothing to write home about when one thinks of our own throwaway economy. It is clear, given the constraints of energy and material resources, that we cannot go on with an ever increasing rate of production and consumption. But to do this would require very different priorities and, above all, a different state of mind.

We don't declare war on famine. We don't work out complex pincer movements on disease and malnutrition. There is no "Stormin' Norman" backed by vast financial resources, leading

The Search for Alternatives

the troops into battle against human suffering and privation. There are no serious debates in our parliaments on how to re-allocate some of the billions spent daily worldwide on armaments to more worthy causes. We can always find billions to develop the newest military technology, but we often literally leave to charity the funding of lifesaving programmes. As an engineer and a scientist, I marvel at what science and technology could do for society. I am, at the same time, horrified at what it is doing. The skills and facilities which, nightly, made possible those military Oscar performances, could be deployed for such humane and creative ends. It is within our grasp to feed the hungry, to make the lame "walk" and the blind "see". That is no longer so much a question of technical ability as one of political will and morality.

DEFINING THE FUTURE

In the absence of mechanisms where communities can be active and people can begin to define their own futures, the real danger is that we shall end up with a very highly automated form of society in which human beings will be passive and the system active. It is very important that we begin to define the future and human-centred technologies are important ingredients. We need the totality of our humanity and not that which is simply made.

One of the major contributions would be for more women to come into science and technology. Not as imitation men, or as honorary men, but to begin to question the value systems being built into our science and technology and show how it can be structured differently.

The most advanced computer-based systems are still trivial compared with the human mind. Apart from its enormous data processing capability it brings with it imagination, consciousness, will, ideology, humour, political aspirations, all those things which make us unique and precious as human beings.

All of our human progress has come about because people have had the courage to question what is going on. It seems to me that, at this stage in our history, we must recognise we are the subjects of history, we are not the objects on which it should wend its painful way. We can and should question what is happening to us, talk, and be willing to stand up and try and do something about it. We must not be bludgeoned into silence by the scale and complexity of the activities and technologies. We must recognise the future is not out there in the sense a coastline might be before someone discovers it. The future hasn't got pre-determined shapes and contours. It has to be built by people like you and me. We do have free choices.

First published:

Technology and International Development
Occasional Paper Series No. 1
Department of Sociology, University College Cork

The Myth of the Moral Neutrality of Technology

First published in *AI&Society* journal, 1995

"Basically, I exploited the phenomenon of the technician's often blind devotion to his task. Because of what appeared to be the moral neutrality of technology, these people were without any scruples about their activities."

Albert Speer, *Inside the Third Reich*

I worked for some 25 years in the aerospace industry. My work was primarily in the design, R&D and test equipment areas. It was and is, an industry employing a very wide spectrum of scientific and technical disciplines and capabilities. Just about every area of scientific and technical endeavour is to be found there in one form or another. Activities ranged from close to basic research through to metal cutting on the shop floor.

It is an industry closely interwoven with the military/industrial complex and although there are many civil products, some 60% of the work is on military equipment including advanced weapon systems. The use or the threat to use such systems raises sharp ethical and moral problems for those who have conceptualised, developed and manufactured them. The collective and individual responses were a revelation. The general context within which the individual justified or explained his or her work is important.

In all my years in that industry, both in a senior design post and

The Search for Alternatives

as a representative of the union which organised the design, scientific and technical staffs, I never ever encountered an individual who said they wished to work on weapon systems. Usually, they simply wished to have interesting, challenging and relatively well paid work and the military industrial complex was the only context in which they could find that.

During the Cold War period, discussion on the morality and ethics within the work context with one's peers was very difficult. Firstly, those who raised these issues tended to be regarded as suspect and guilty of undermining the national defence efforts. The existence of the Official Secrets Act during this period essentially silenced such discussion at work, and rendered it positively dangerous with outside friends and professional colleagues.

Scientists and engineers have failed in all countries to develop professional bodies which embody and safeguard ethical codes and from whom the individual scientist or engineer may seek advice, guidance and support. There is no equivalent for engineers and scientists of the ethics committee to which those in the medical field may turn. There exists no mechanism to even depersonalise an ethical conflict with an employing body by asserting "I cannot undertake that as it is contrary to the ethical requirements of my professional body". Engineers and scientists do not have a Hippocratic Oath although Professor Thring did attempt to gain support for one in the 1960s and 1970s. More recently in Germany, professionals in management and marketing have been supporting a Hippocratic Oath.

Whilst not wishing to overstate the significance of a Hippocratic Oath, it does at least provide some collective professional benchmarks against which actions may be judged.

Historically, the scientist and engineer has had to address these issues in the context of individual conscience. That is as it should be at a first level but the absence of a collective to which the individual can appeal for support, leaves the individual in an isolated and sometimes defenceless position. Even where collective bodies of scientists are formed, specifically to encourage a sense of responsibility, that is interpreted as heightening the responsibility of the individual (laudable though that is) rather than providing a framework of collective action and support.

The British Society for Social Responsibility in Science (BSSRS) which held its inaugural meeting at the Royal Society in April 1969, did provide a stimulus and context for the discussion on social responsibility. It did not however, succeed in translating this into a context of collective support although it should be pointed out that when the author, who was a member of BSSRS, was sacked by Lucas Aerospace, letters of protest and support were organised by Joseph Needham, Steven and Hilary Rose, Richard Fletcher and a number of MPs. Letters to the company poured in from scientists and politicians worldwide – 148 German academics signed letters of protest. When the company refused to change its mind and a strike ensued, demonstrations outside the plant were addressed by BSSRS members including Maurice Wilkins (of DNA fame) and Jonathan Rosenhead. However, even this did not save my job.

Cynics were quick to point out that scientists who had literally changed the course of history, and even some strong, well organised local trade union groups could not defend a colleague. However, that would be to ignore the consciousness raising activities of such events. The action helped to put the notion of socially useful technology and environmentally desirable products on the political agenda at a time when it had yet to become fashionable. It is in that wider context that the aspirations of BSSRS and its predecessor in America can best be understood.

> "To foster throughout the world a ... tradition of personal moral responsibility for the consequences to humanity of professional activity, with emphasis on constructive alternatives to militarism; to embody in this tradition the principle that the individual must abstain from destructive work and devote himself to constructive work, according to his own moral judgement; to ascertain the boundary between constructive and destructive work, to serve as a guide for individual and group decisions and action…"

For many people, accepting the job is the point at which they

The Search for Alternatives

exercise their moral judgement. You know what a company is doing, you know what experiments are going on in a particular lab and if you decide to work there, you accept all that lock, stock and barrel.

For some, the division between professional and citizen proves attractive. They do one thing as a professional in their job and then exercise their moral judgement and social conscience by their out-of-work activities, the churches to which they belong and the political parties they support. Others will assert that it is up to society to decide how particular areas of science are used and how technology is applied. This they see as a separate political dilemma not a professional one.

Some clear their conscience by the 'drug dealer' explanation: 'If I didn't do it, somebody else would'. Some will assume a sort of Eichmann role: 'I'm only doing what I was told'. Others feel the moral dilemma is too great and simply resign. My own view is that an employment contract cannot and should not suspend one's ethical and moral concerns. In fact I hold that we have a responsibility to seek to address these issues within the context of our professional and occupational situations.

The issues described above are particular to the defence industries but variations of them occur in areas of biology related to warfare, bio-engineering, animal experiments and others. Those engaged in fundamental research assert with justification that in their area it is not clear in any case where their work will lead and the pursuit of knowledge should not be impeded. However, it should be pointed out in passing that the form of project which excites the scientific mind and attracts funding inevitably displays sets of priorities and values which are deeply rooted in the Western scientific tradition.

It may be helpful to illustrate some of these issues by considering a specific area, e.g. the introduction of new technologies and the impact on those who use them and more broadly, the impact on the environment. Attempts have been made to come up with a set of values which might apply in such circumstances. An interesting example is that of the International

The Myth of the Moral Neutrality of Technology

Association of Machinists' Technology Bill of Rights.

On April 30 and May 1 1981, William Wimpisinger, then President of the Association and Aerospace Workers hosted the International Association of Machinists scientists and engineers conference in New York. It was chaired by Seymour Melman, Professor of Engineering and Operations Research at Columbia University. The Technology Bill of Rights was produced as a direct result of the conference. Some of its provisions were as follows:

> Para 1. 'New Technology shall be used in a way that creates jobs and promotes community-wide and national full employment…'
> Para 4. 'New Technology shall improve the conditions of work and shall enhance and expand the opportunities for knowledge, skill and compensation of workers. Displaced workers shall be entitled to training, re-training and subsequent job placement or re-employment.'
> Para 10. 'When new technology is employed in the production of military goods and services, workers through their trade unions and bargaining agents have a right to bargain with the management for the establishment of alternative production committees which shall find ways to adapt the technology to socially useful production in the civilian sector of the economy.'

The Lucas workers, likewise sought to introduce a socially responsible framework in which technology and new technologies in particular might be applied. They created a framework with twenty attributes which included the following:

> 1. The process by which the product is identified and designed is itself an important part of the total process.
> 2. The means by which it is produced, used and repaired should be non-alienating.
> 3. The nature of the product should be such as to render it as visible and understandable as possible and compatible with its performance requirements.
> 4. The product should be designed in such a way as to make it repairable.
> 5. The process of manufacture, use and repair should be such as to

The Search for Alternatives

conserve energy and materials.

6. The manufacturing process, the manner in which the product is used and the form of its repair and final disposal should be ecologically desirable and sustainable.

7. Products should be considered for their long-term characteristics rather than short-term ones.

8. The nature of the products and their means of production should be such as to help and liberate human beings rather than constrain, control and physically or mentally damage them.

9. The production should assist cooperation between people as producers and consumers, and between nation states, rather than induce primitive competition.

10. Simple, safe, robust design should be regarded as a virtue rather than complex 'brittle' systems.

11. The product and processes should be such that they can be controlled by human beings rather than the reverse.

12. The product and processes should be regarded as important more in respect of their use value than their exchange value.

13. The products should be such as to assist minorities, disadvantaged groups and those materially and otherwise deprived.

14. Products for the Third World which provide for mutually non-exploitive relationships with the developed countries are to be advocated.

15. Products and processes should be regarded as part of culture, and as such meet the cultural, historical and other requirements of those who will build and use them.

16. In the manufacture of products, and in their use and repair, one should be concerned not merely with production but with the reproduction of knowledge and competence.

When a general framework of this kind has been agreed, it can give rise to a rich philosophical and practical discussion as to how these objectives can be transformed into design and engineering practice. In considering 6 above, "Ecologically desirable" raises profound issues about how we relate to nature and how we use the natural resources around us. It caused us to perceive ourselves as being part of nature and gradually to understand and accept that if we damaged the natural world around us we were also damaging ourselves. You cannot have healthy people on a sick planet. It

The Myth of the Moral Neutrality of Technology

opened up our ever focussed vision to insights from other cultures, e.g. "Man did not weave the web of life, he is merely a strand in it. Whatever he does to the web he does to himself" (Chief Seattle). It stimulated a discussion on science and technology which is today so ideology-laden and so integrated into our scientific outlook that expressions such as "the manipulation of data", "the exploitation of natural resources" and "the control of nature" were not considered controversial.

Translated into practical design, these considerations led directly to the design and testing of a long-life repairable, low environmental impact car. Indirectly they led to the design of a new form of permanent magnet motor which significantly reduced energy consumption and in addition to a range of energy conservation devices and alternative energy systems including heat pumps using natural gas. Products such as these and many other reflecting what might be called a new engineering ethic, are to be found in the Product Bank of the Technology Exchange. They also introduce a new ethical concern about sustainability.

For readers of this journal, it will be those issues relating to information technology which will be of great interest. It followed from 8 above that human skill, ingenuity, creativity, imagination and purpose would be perceived as an asset rather than a liability. To accept this at a subjective level and also in the context of design methodology is more radical than may at first appear. The Taylor philosophy, "In my system the workman is told precisely what he is to do and how he is to do it and any improvement he makes upon the instructions given to him is fatal to success", is clearly out. We would no longer refer to the human brain as "the only computer built by amateurs". Nor would we comport ourselves as to accept a future in which "human beings will have to accept their true place in the evolutionary hierarchy – animals, humans and intelligent machines".

As we embarked upon the design of complex systems involving human beings and machines, we would set about ensuring that this was done in the design context which allowed the human being to give vent to his or her creativity. We would not so objectivise

The Search for Alternatives

human knowledge as to render the human being passive and the systems active thereby solely conferring life on machines and diminishing human beings. We would have to conceptualise systems such as tools in the Heidegger sense rather than machines. It would follow too that we would attempt to design systems which are transparent rather than opaque.

It would and did cause us to re-examine the significance of Durer's advocacy of "instruments of labour through the use of which the conceptual part of work will be made manifest" and "the development of new forms of mathematics which would be as amenable to the human spirit as natural language".

Thus a simple ethical position of respecting human beings and their skill and ingenuity, resulted in profound design considerations. In translating these ideas into practice, the Lucas workers initially advocated forms of telechiric devices which would have audio-visual and *tactile* feedback and would respond to human skill and decision making without objectivising that knowledge. These consideratons in turn led to proposals for human centred systems and indirectly to EEC projects such as ESPRIT 1217 "to design and build a human centred advanced manufacturing system".

Starting from a quite straightforward ethical concern for human beings, an alternative way of viewing the design of advanced systems began to emerge and resulted in significant questioning of many given assumptions, not least on what constitutes a scientific design. We had come to believe that a system is only scientifically designed if it displays the three predominant characteristics of the natural sciences – that is to say predictability, repeatability and mathematical quantifiability. It follows by definition that this precludes intuition, subjective judgement, tacit knowledge, heuristics and imagination which are all distinguishing features of our humanity. Thus the natural sciences may be perceived to be particularly unnatural.

It follows that if we proceed on the basis of scientific methodology based on causality, we are inevitably drawn to the conclusion that the human being is a machine. However, if we

The Myth of the Moral Neutrality of Technology

proceed on the basis of an explanation based on purpose, we would at least have an equivalence of explanation which opens up fascinating possibilities.

It would follow from these considerations that we would re-examine concepts such as training. In its modern form training normally connotates narrow machine specific competence. It is very much within the Tayloristic tradition. Interestingly, no words for training exist in the German or French language, for example, and Japanese companies refer to 'on the job learning'. I frequently point out that one programmes a robot, trains a dog, but educates human beings. Apprenticeships in the classical sense were actually the transmission of a culture as much as providing for the basis of manual dexterity and diagnostic capabilities. It follows from this that knowledge updating and multimedia learning systems would not seek to transmit expertise but rather involve the learner in a process through which expertise is acquired. They would be to the learner somewhat like a flight simulator is to the pilot. They would bring you closer to the point at which you could begin meaningfully to deal with "the real thing".

These considerations would lead us to design systems in the tacit area of knowledge rather than at the data level where the noise is high and the signal is low. Interestingly, such concerns were so poetically described by TS Elliot. Indeed, frequently the big issues in science and technology are anticipated by artists, poets and novelists and the professional ethic which in practice regards these as diversions from "reality" reminds us of the 1960s slogan: "Do not adjust your mind there is a fault in reality". That reality has now so monopolised the scientific and technocratic mind to such an extent that in one leading university in the USA scientific and engineering students are specifically recommended to avoid reading fiction since by definition fiction dealt with the unreal and as engineers they had to prepare themselves for a world of reality and such reading would clearly be a serious diversion.

It is my hope that as the year 2000 approaches it will provide a stimulus to re-examine the ethical and moral values deeply rooted in our professional activities. The sheer rate of change should

The Search for Alternatives

cause us to reflect carefully upon what it is we are doing to the flora and fauna around us and to ourselves. In the last 100 years alone we learned to fly, killed 17 million people in one war, declared Jackson Pollock to be a great artist and played golf on the moon.

Science and technology are now the leading edges in Western society as religion might have been at earlier historical stages. To cope with this changing situation, we need to encourage more widely across society and in science and engineering in particular, a more holistic approach which encourages the perspective of a historian, the wisdom and care of nature of a Chief Seattle and the analytical power of the scientist.

We are the newest and most immature of the species around us yet we stand as the first generation of the only species that ever had it within its power to destroy the earth and humanity as we know them. We urgently require a philosophical, cultural and ethical outlook which will equip us to deal wisely with our awesome new found power. My own view is that we have become far too smart scientifically to survive much longer without wisdom.

These are issues of great complexity. They do not yield up a straightforward right or wrong as a mathematical formula might do. "Thou shalt not kill" can so easily transform into "Thou shalt not kill unjustly". These dilemmas have haunted those of conscience engaged in scientific and technological endeavour throughout our recorded history. Leonardo decided he would keep secret his inventions fearing the terrible uses to which they would be put by "the perversity of man". However, when devising means of mass destruction of his employer's enemies he wrote:

> "When besieged by ambitious tyrants I find a means of offence and defence to preserve the chief gift of nature which is liberty."

Nor does there seem to be a great contradiction in using religious metaphors:

The Myth of the Moral Neutrality of Technology

> "It is the most deadly machine that exists. And when the cannon ball falls, the nucleus sets fire to the other cannon balls and the central ball explodes and shatters the others which catch fire in the time it takes to say a Hail Mary."

The massive ethical contradictions between means and ends is vividly encapsulated in the statement attributed to Alfred Nobel:

> "I would like to produce a substance or machine of such frightful, enormous, devastating effect that wars would become altogether impossible."

Scientists and engineers, in their day to day work, have both a personal and professional responsibility to consider the ethical basis of the work they undertake. However, once they identify an issue, there is a glaring lack of framework or context in which to discuss issues. This deficiency is now being recognised and 'Scientists for Social Responsibility' have embarked upon an ethics initiative. They are lobbying for the setting up of a nationwide structure of ethical committees in all universities in the UK. Likewise, they propose setting up a counselling network as a support service for anybody experiencing ethical difficulties arising from their employment. In a quite imaginative and practical move, they are setting up a compendium of scientists including some eminent figures willing to give young scientists and engineers free advice about their career choices based on their own experiences from a standpoint of ethical and scientific responsibility.

It is to be hoped that scientists and engineers in the areas of new technology including artificial intelligence, will participate in movements of this kind. It was also encouraging to see that at the SSR (Scientists for Social Responsibility) Annual Conference in London in November 1993 the trade union MSF was a co-organiser of the event.

As the 20th century draws precariously to its close and as the multiplier effects of our science and technology become more and more visible, it is to be hoped that a culture will be imbued among scientists and engineers in which ethical questions will be regarded as pivotal rather than a diversion from getting on with 'scientific progress'. An important start would be to include such topics in undergraduate engineering and science courses.

First published:
AI & SOCIETY, VOL. 9:10-17 (1995)

MY EDUCATION

First published in *My Education*, 1997

I think that the most important influences for me were all outside of school. Schools were unstructured, they were often dangerous. Somebody once said that education is the structure on which you hang the rest of your life, but, if that were true, I would think of it as a wall in a gallery. It is flat, it is monotonous, it has a top, a bottom and two sides. Using the analogy of an art gallery, it really only becomes interesting when you begin to paint pictures which you hang on it.

Tuam in the 1940s was a typical provincial town, in a number of ways. It had the advantage that there was the sugar factory, so there was a sense of an emerging technology, as it would have been at the time. The outer parts of the town still didn't have electricity, so I was able to see at first hand the transformations which a new technology can make, both positive and negative, on the lives of people. And there were extraordinary people in the town. There was an exceptional range of skills of all kinds. Many people in the town were, in my view, unselfconscious artists. The manner in which women could metamorphose a wedding dress into a Confirmation dress for a daughter and again into a Holy Communion dress seemed to me an astonishing skill.

This influenced my subsequent thinking, because I now refer to technologies where you have a 'cascade use' of the material,

The Search for Alternatives

using the same piece of material in different roles over a period of time. I got those deep insights in Tuam at a very early age. There were blacksmiths in the town; John Connolly was one and he used to draw the gates he was going to make in the ashes with the tip of the poker and say to the farmer, 'Would you like a twirl on it here?' referring to the metal turns on the gates. As a result of all these different skills, I was very conscious of the need to preserve them and build on them and this advised a lot of the work I did subsequently on human-centred systems. This tacit knowledge has been a great area of interest of mine.

Tacit knowledge was explained by the philosopher of science, Polanyi, who said that there are things we know but cannot tell. It is a sense we have of shape, size, form and appropriateness which we acquire through practice, through relating to materials, to working with materials. Since it cannot be written down or explicitly stated, there is a tendency for modern educational systems to ignore it or say that it doesn't exist, that the only important things are those things which we can state explicitly. I saw vivid examples of this during my childhood. There was a stonecutter in the town who made a gravestone for our family and I remember him saying to me, 'If you come back next week the head will be coming out of the stone', as though it had been born from the stone. He was really like Michelangelo; he could already see the figure in the material and all he had to do was remove all that which was not the figure.

I remember going to Galway when they still used to make the Claddagh shawls. It was an astonishing piece of textile design, in the best sense of the word. I have seen people get degrees in fine art at the Royal College of Art in London who, in my view, couldn't even begin to approach those kind of skills although they could write about them – so I became conscious very early on that our society values linguistic ability more than real intelligence and that is something I have tried to address throughout my lifetime.

The art of storytelling had an enormous influence on me. People used to gather in particular houses and they would tell stylised versions of the great old classical stories from the West of Ireland.

My Education

I still have childhood images of these Rembrandt-like heads who would vividly portray a story. As children, we tried to emulate this. We used to have games where we would try to tell a story that would make somebody laugh or make them cry. I used to succeed, on occasion, in making the other kids cry with my very moving story. This convinced me that there was a real possibility of having a whole range of 'magic carpets' in one's mind. When I subsequently read about the Arabian Nights, I became aware that you could conjure up images of imaginary countries, where people cared for nature or were kind to each other, or of extraordinary machines which allowed human beings to do all kinds of things. So, through that story-telling tradition, my imagination was greatly inflamed and I still have a reservoir of magic carpets on which I can fly whenever I am moved to do so.

Playing hurling taught me more than just the game itself. Each road would play against the others and I think there were seven roads in the town. I was conscious very early on that how you defined a road often determined whether you stood a good chance of winning. If all your best players were in a road within the precincts of the town, you would try to get an agreement early on that it was only the roads up to the edge of the town that counted. We had a number of players who lived out on the Milltown Road and they were excellent players, so we wondered how we could somehow change the rules. We decided that a road would be defined as a road through which you went on your way to school, which meant that we had a much bigger catchment area. A year or two afterwards, if we found that others were using the catchment area, we might try and change the ground rules again. A lot of my capacity to negotiate in trade union work or even in complicated international contracts was established then, so, even in childhood games, one can learn an enormous amount.

I liked languages very much. I always point out that I learned German rather than studied it. As children, we used to go a lot to the cinema – the 'fourpenny rush' – on a Sunday afternoon. There were still films of the war and cuts of Hitler. I was fascinated by the resonance and dynamism of his speeches, without having the

The Search for Alternatives

slightest idea of what they were about. I subsequently had a colleague on the Ballygaddy Road and I used to go and listen to his records, which were old 78s of Beethoven and Mozart and so on. It really did jar on me that a nation that had produced the beauty of Beethoven and Mozart could also produce the hideousness of Hitler. I felt I should try and find out something about that language and that culture. I went down to the Kaplan family – he was the chief engineer in the sugar factory – and I announced to Mrs Kaplan that I wanted to come and learn German. When she got over her shock, she agreed to this and, by the time I was eighteen, I could speak fluent German. I now present television programmes in German and I write books in German, but I have never formally studied German. That seems to me to say a lot about the educational process. One needs the excitement, the motivation and then the capacity to draw on resources around one. Even in Tuam, which was not a centre of Germanic study, if one used the imagination, it was possible to find suitable resources.

I was very very keen to learn workshop technology and have access to a lathe and that type of machine tool and also to learn technical drawing. It seemed to me to be a marvellous subject, but it was not available in the Christian Brothers school. I asked if I could have Wednesday afternoons off, because I had worked out that this was the time it was being taught in the 'tech' [Technical School]. The Christian Brothers were really horrified that anybody should want to go to the tech and they wouldn't agree to it. I then spoke to my parents about it and they agreed that I could continue Latin and German and so on at the tech full-time. I still remember the shock on his face when I told Brother Rafferty. 'If you go to that place' – he didn't even call it a school – 'if you go to that place, you will be finished,' he said. That showed me that, often, educationalists and others try to protect you from what they regard as dangerous or uncertain or unpredictable, when it is precisely through such a thing that we often gain the richest insights. So I shaped my own comprehensive education.

In the tech I met an extraordinary teacher, Sean Cleary. I am sometimes very critical of formal education but I always highlight the marvellous exceptions, the great gifted teachers I met, and he was one. He had worked in England for Vickers and elsewhere, so he could give you a vision of what engineering and metalwork could be like. He also had a deep sense of quality of workmanship and he would sometimes hold up a piece of material and say, 'Isn't that beautiful, just surface finished?' The culture that was transmitted just by doing that was very powerful. I told him I was very keen to design and build a steam engine and he was a good enough facilitator and teacher to say, 'Well, alright, let's try and do it.' The only machine tool we had was a lathe and with him I designed and built a double-acting steam engine which I have to this day. It was a very, very formative experience. We needed close-grain cast iron for the cylinders and there was simply none in Tuam. Now if I was going to be a good civil servant I would have learned at that stage that you write a report about the material you want, you then write saying there is none available and you conclude that the project can't be done. But we heard that there was an old sawmill with an abandoned fly-wheel and we knew that a fly-wheel has to stand high centrifugal forces, so it would have been made from good close-grain cast iron. So we went and cut a huge chunk of cast iron out of it, divided it up into pieces and made a marvellous steam engine.

I think the education system prepares people to analyse rather than to do. The great strength of a craft tradition articulated through a sensitive education system is that it encourages this sort of can-do mentality. I went to work in the sugar factory for about nine months before I went to Germany and Switzerland to study and there, if a great machine broke down, the whole place was geared to getting it working again. They didn't spend hours and hours talking about the things that could not be done and that for me was a tremendous reinforcement of a positive can-do mentality, rather than a can-analyse mentality. People often mistake apprenticeships for simply the transmission of manual dexterity, whereas it is actually the transmission of a great culture, of how to organise yourself, how to get materials, how to plan things. It was in the sugar factory that I also met other apprentices. There was

The Search for Alternatives

Michael Hussey, who has done all kinds of extraordinary things in the meantime. Tom Murphy was also in the factory at that time, making up poems and plays, and Michael Brennan, who is now a trade union leader. The very first day I went into the factory, Michael came up and to me and said, 'Would you give me a bob?' because he wanted to send a telegram to Castro to congratulate him on arriving back in Cuba. This period was politically and technically formative for me.

The Kaplan family suggested that I go to the Continent and study engineering and German, which indeed I did. There were no agencies that would arrange exchanges then, so I had to write to different embassies finding out how I could go and how to get a passport and where I could stay when I got there. Eventually, I made the necessary arrangements and went abroad to Switzerland first and then to Germany and it proved to be an extraordinary educational experience in the truest sense of the word.

I worked with Lucas Aerospace in England for some eighteen years. We were working with very advanced technology – Concorde and fighter aircraft and so on. It seemed awful to us that there was all this human suffering around when technology properly applied could do so much to alleviate it. We were working on advanced guidance systems that can guide missiles to another continent with extraordinary accuracy, but the blind and the disabled were still staggering around as they had in medieval times. So, rather than accept structural unemployment, which was then being proposed in the company, our view was that all that skill and talent should be used to improve the quality of life for people. We came up with an extraordinary plan for socially useful production. About one hundred and fifty products would be made which would reduce energy consumption, would dramatically reduce pollution and would make meaningful creative jobs for people. Although we were unsuccessful, in the sense that I was sacked in a big blowup amid worldwide protest, the ideas were broadly correct. I think the ideas are as relevant today as they were then.

In a strange way, artists and poets often prefigure the really big issues in society and we as engineers and scientists diminish ourselves if we are not exposed to those ideas. That is expressed for me most powerfully in the part of *Finnegans Wake* where Joyce describes the two opposites that make up each person: Shem and Shaun, the positive and the negative. Of those who always emphasise the negative, he says, 'Sniffer of carrion, premature gravedigger, seeker of the nest of evil in the bosom of good word. You who sleep at our vigil and fast at our feast, you with your dislocated reason, you have reared your disunited kingdom on the vacuum of your own most intensely doubtful soul.' That for me highlights a big issue in design. Do we design systems that assume all the things people can't do (where they talk about fool-proof systems, they actually mean that the people are fools), or do we look at all the greatness, talents and abilities of people and, instead of designing systems to obliterate that, to reduce human beings to abject, pathetic machine appendages, actually enhance those skills and abilities? I got that insight from *Finnegans Wake* and it is now an area of research and design methodology which is being pursued worldwide.

 I was conscious very early on of what a natural treasure trove I had all around me and I really relished that, in a childlike way of course. I used to marvel when the skeins of geese would return in October; it was always the third week in October and I would wonder how they had guided themselves, whether they were using the earth's magnetic lines or the stars. That was later to excite my interest in guidance control systems. I sometimes used to spend the summers in Galway with my uncle Martin. Looking down at the salmon weir. I was fascinated by the fact that salmon could find their way back to the same river in the same part of the world, having been thousands and thousands of miles away. These astonishing features of nature really did awaken my appetite and I still enjoy nature enormously and have tremendous respect for it.

In recent years, I came across the work of Chief Seattle, which encapsulates for me many of the feelings I myself used to have about nature in the West of Ireland. I never felt separate from nature there, I always felt a part of it. I felt as much a part of it as

The Search for Alternatives

the curlew or the beautiful cloudscapes we used to have – rain was never offensive to me. I enjoyed the whole surroundings and Chief Seattle for me expresses that feeling of closeness to nature and then presents a sort of religion and outlook which separate us from it and put us totally above it. He confronted one of the American presidents about the destruction of the buffalo and the president said, 'You're going to have our culture, our language, our literature and our religion.' Chief Seattle responded in the most extraordinary fashion, when he said, 'Every part of this country is sacred to my people. Every hillside, every valley, every plain and every grove is hallowed by the memory and experience of my tribe. Even the rocks in the sea are charged with our memories. The dust under your feet responds more lovingly to our footsteps than to yours, for the soil is rich with the life of my people. Our religion is the tradition of our ancestors and is written on the hearts of our people. Your religion was written on tablets of stone by the iron finger of an angry god.'

That seems to me to be profoundly insightful and, as we approach the twenty-first century, I think it should create a psychological stimulus to make us re-examine the strange, double-edged journey which has brought us to where we are.

First published in:
My Education
John Quinn in interview with..., Town House, 1997

From Judgement to Calculation

First published in *AI&Society*, 2007

IT systems frequently come between the professional and the primary task as the real world of touch, shape, size, form (and smell) is replaced by an image on a screen or a stream of data or calculation outputs. This can lead to high levels of abstraction where the ability to judge is diminished. I have described elsewhere the case of a designer using an advanced CAD system who input the decimal point one place to the right and downloaded the resultant output to the production department on a computer-to-computer basis (Cooley 1991). The seriousness of this error was further exacerbated when the designer, shown the resulting component which had been produced, did not even recognise that its dimensions were ten times too large.

Scientific knowledge and mathematical analysis enter into engineering in an indispensable way and their role will continue. However, engineering contains elements of experience and judgment, regard for social considerations and the most effective way of using human labour. These partly embody knowledge which has not been reduced to exact and mathematical form. "They also embody value judgments which are not amenable to the scientific method." (Rosenbrock 1977).

These will be significant issues as IT is increasingly deployed in societal areas such as that of healthcare. Cases already abound and many have become high profile public issues, e.g. the paediatricians who administered a fatal dose of 15 mg of morphine

The Search for Alternatives

instead of the correct 0.15 mg for the baby (Rogers 1999; Joseph 1999). They did this in spite of being warned by a staff nurse that the dose was obviously incorrect.

Those introducing the avalanche of new technologies frequently limit their considerations to first order outcomes. These usually declare the positive and beneficial features, whilst only fleeting attention is given to the downside, if at all. It is as if the laws of thermodynamics no longer apply and that you can get something for nothing. We are now beginning to learn, to our cost, that there are "no free dinners" with technology. For too long we have ignored the double edged nature of science and technology (S&T). Viewed in this light, it has produced the beauty of the Taj Mahal and the hideousness of Chernobyl, the caring therapy of Röntgen's Xrays and the destruction of Hiroshima, the musical delights of Mozart and the stench of Bergen Belsen.

Most technologies display positive and negative aspects. There is now an urgent need for a new category of competence – an ability to discern the positive and negative aspects of a given technology and to build upon the positive whilst mitigating the negative features. It is not a question of being for, or against technology but rather discerning the positive and beneficial uses of it.

One negative aspect of IT technology is the under-valorisation and frequently the squandering of our society's most precious asset which is the creativity, skill and commitment of its people. Over the past 21 years *AI and Society* has facilitated a debate on positive alternatives to the existing developments and has placed particular emphasis on the potential for human centred systems. Its articles, reports and the conferences it has facilitated have provided practical examples and case studies of systems design which celebrate human talents.

It requires courage, tenacity and profound insights to develop these alternatives in our obsessively technocratic and machine centred culture.

From judgement to calculation

THE WOW FACTOR

Technology in its multi-various forms is rapidly becoming all pervasive. It permeates just about every aspect of what we do and who we are. It ranges from the gigantic, such as the diversion of rivers and the repositioning of mountains to the microscopic level of genetic engineering. Science fiction becomes reality as faces are transplanted and head transplants are confidently predicted.

The "wow!" factor is mind-blowing. Even simple internet procedures have a God-like quality. With Google Globe you can look down on our planet and travel over continents and countries, quickly homing in on an aerial view of your beloved "homestead" showing your own car in the drive.

We now appear as masters of the universe, able to see everything and confident in the belief that any problem we create we can also solve. It is just a question of a plentiful flow of research grants and resources. Meantime, we plan to bury our nuclear waste.

AWESOME CAPABILITY

We are the only species ever to have it within its power to destroy itself along with our beautiful and frail planet. This is an awesome capability and one for which our culture, education and politics ill prepares us to cope creatively. Change is frequently and thoughtlessly portrayed as progress and progress so unidimensionally defined is evident on all sides.

In spite of this, at no time in history have so many people been fearful of the developments surrounding them and are becoming alienated from the society producing them. Doubts are jolted into concerns by global warming events or the looming spectre of an Avian Flu pandemic. Yet it tends to be a fear that dare not speak its name. Who, after all, can be against progress, even if it is defined in its own self serving terms?

The Search for Alternatives

PATHS NOT TAKEN

In order to analyse where we are now with IT systems it is important to look back historically to identify turning points at which technology might have and could have developed differently. This is akin to Rosenbrock's notion of the 'Lushai Hills Effect' (Rosenbrock 1988, 1990). He suggests that with technology, we sometimes take a particular route of development and once we have done so we begin to believe that it is the only one. We then develop cultural forms, educational systems and a philosophical outlook which supports that contention. It therefore seems useful at this juncture to explore different interpretations of human and technological progress which may throw light on our present dilemma and indicate alternatives worthy of exploration.

EGO SMASHING EVENTS

We are indebted to Mazlish (1967) for the notion of technological and scientific development as dismantling discontinuities in historical ego smashing events. The first arises from Copernicus and Galileo which resulted in a re-organisation of the universe with our earth no longer at its centre. The second is based on Darwin who robbed human beings of the particular privilege of having been specially created. The third, based on Freudian insights, suggests that we are not the masters of our own consciousness in the way we had assumed ourselves to be. Our society is now apparently demolishing the fourth discontinuity – the one between humans and their machines.

SELF ELIMINATION

> "To put it bluntly, we are now coming to realise that man and the machines he has created are continuous and that the same conceptual systems that help to explain the workings of the human brain also explain the workings of a thinking machine. Man's pride and his refusal to acknowledge this continuity is the sub-stratum upon which the

From judgement to calculation

distrust of technology and industrial society has been reared"(Mazlish 1967).

However, as we shall suggest later, this sub-stratum of distrust may be overcome if we view human beings and their machines as constituting a symbiosis rather than a convergence. Otherwise, as Karl Pearson (cited in Weizenbaum 1976) puts it: "The scientific man has above all things, to strive at self elimination in his judgments" (Pearson 1976).

WALKING, FEEDING, THINKING

Another conceptual framework which yields interesting insights is to consider technological change as a series of phyla. Rapoport (1963) identifies four.

The first phylum consists of tools. Tools appear functionally as extensions of our limbs. While some mechanical advantage may be gained from such a device, it in no way functions "independently of us."

The second phylum is mechanical "clockworks." Here the human effort in winding up the mechanism is stored as potential energy which may be released.

Over a long period of time the clockwork gives the impression of autonomous activity. Furthermore, it is not a prosthetic device to extend our human capabilities but rather one that produces time: hours, minutes ... to pico-seconds. Thus in his seminal work, Lewis Mumford asserts that it is the clock and not the steam engine that is "the key machine of the modern age" as it "dissociated time from human events and helped create the belief in an independent world of mathematically measurable sequences: the special world of science" (Mumford 1963).

Weizenbaum points out that clocks "are the first autonomous machines built by man and until the advent of the computer they remained the only truly important ones." He also asserts "This rejection of direct experience was to become one of the principal characteristics of modern science" (Weizenbaum 1976).

The Search for Alternatives

The third phylum is heat engines. These gradually emerged as devices that were neither pushed nor pulled but "fed." The fourth phylum covers devices capable of collecting, storing, transmitting, manipulating, initiating information and determining actions based on these.

It will be seen that in each phylum, the device moves toward autonomous capabilities but there is also a form of narcissism – technological narcissism – as clockworks "walk", heat engines "feed" and computers "think." We design devices with some human attributes and then in a strange dialectical way we begin to perceive ourselves as partial mirror images of the machines. During the early stages of clockworks, drawings showed human sinews and muscles in machine-like manner and Déscartes refers to the human being as a machine. In the era of heat engines there is a growing concern about what and how humans are fed. This is sometimes reflected in concerns about dietary intake and some even suggest could lead to anorexia.

The fourth phylum leads to a situation where someone could say disparagingly "The human mind is the only computer made by amateurs" and a high priest of technology was presumably half joking when he said "Human beings will have to accept their true place in the evolutionary hierarchy: animals, human beings and intelligent machines."

FAULT IN REALITY

The foregoing provides an interesting context in which to view the potential for human centred systems. However, the discussion of such systems has suffered from its questioning of the given orthodoxy in contemporary science. To do so is to elicit the disapproval of many of one's colleagues. Sympathetic colleagues may imply that you have not grasped the greatness of all that is going on. Less sympathetic colleagues hint that you are questioning rationality itself and are therefore guilty of irrationality.

Although Stalinistic psychiatric wards are not threatened, grants

From judgement to calculation

may dry up and you can forget that tenured post. Perhaps the students in the sixties had a point with their posters: "Don't adjust your mind. There's a fault in reality."

Our culture conveys the sense that a calculation is precise, analytical and scientific. It is regarded as apolitical and objective. Indeed in the sixties, when social scientists were struggling to gain acceptance of their science, many of their papers were awash with calculations and diagrams. However, when I worked in the aerospace industry I found that those who could make best use of computers and calculations already knew in a "ball park" sense what the answer should be and they used computer based calculation as a fine tuning device. They were able to rely on their judgment, so if a discrepancy arose the problem would be revisited.

In spite of this, judgment tends to be regarded as something much less significant. An informed guess – or worse a shot in the dark – is often dismissed as mere speculation. At the level of proficiency, Dreyfus refers to it as "holistic similarity recognition" and points out that "intuition is the product of deep situational involvement and recognition of similarity." This becomes expertise when "not only situations but also associated decisions are intuitively understood" (Dreyfus and Dreyfus 1986). Using still more intuitive skills the expert can cope with uncertainties and unforeseen or critical situations and has the ability to override or disagree with calculated solutions.

Decision making is probably at its best when there is a creative interaction between judgement and calculation. Both have their place in the symbiosis.

INTIMIDATION

Pivotal to all of this must be whether the output of a calculation is correct and how we can verify its status. Calculations, at least in the temporary sense, can be quite intimidating even if they are completely wrong. Archbishop Ussher, in calculating the age of the world as understood in the Middle Ages, declared it was

The Search for Alternatives

created in 4004 BC on October 22nd at about 6.00pm (Ussher cited in Rosenbrock 2002). Although his calculation was wrong by some billions of years it must have seemed quite impressive at the time.

Recently, in a widely publicised trial, the expert witness Sir Roy Meadows declared the probability of two natural unexplained cot deaths occurring in a family was 73 million to 1. The court was impressed. Only later, when the odds were shown to be closer to two hundred to one was the enormity of the error exposed.

I have described elsewhere the shift from judgment to calculation with some of the consequences. Initially, these were in the engineering field but are increasingly occurring elsewhere, e.g. in the medical field. I have represented this graphically as a shift from judgment to calculation; from the subjective to the objective and from signal to noise (Cooley 2002).

The question may arise as to whether this matters significantly. Perhaps the problems identified are merely transitional ones which occur as the systems are being bedded down. It will be argued by many that this is in the nature of the human progress project.

After all, we extended the capacity of our hands through a variety of tools. With spectacles, telescopes, microscopes and scanners we extend our vision. IT technology is merely a further development in which we now extend the capacity of our minds. This is a part of human progress – a speeded up version of the strongest of the tribe climbing to the top of the hill to see what is on the other side. If it could be done then do it!

CAN WE, SHOULD WE?

I hold that it is no longer adequate to ask "Can we do it?" Rather we need to enquire "Should we do it?" The fourth phyla is of a different order to the previous three.

The new technologies under consideration have been developed by appropriating human intelligence and objectivising it into computer based programmes and technological procedures. However, this is becoming qualitatively different from previous technological developments in that more and more humans – even

From judgement to calculation

at the highest professional levels – are becoming increasingly dependent on calculations and systems output.

The deep problem arises when human abilities and judgments so atrophy that we are incapable of disagreeing with, questioning or modifying a systems output. A simple example of this is the increasing number of people unable to add a column of figures, even to get an approximate total.

LOSS OF NERVE

I do believe that we are now at a historical turning point where decisions we make in respect of new technology will have a profound effect upon the manner in which our species develops. As matters now stand we are becoming increasingly dependent – some would say abjectly so – upon machines.

Rosenbrock has cautioned against this approach. In the field of computer aided design, the computer is increasingly becoming a sort of automated design manual leaving only minor choices to the design engineer. This he suggests "seems to me to represent a loss of nerve, a loss of belief in human ability and a further unthinking application of the doctrine of the division of labour." He further points out that the designer is thus reduced to making a series of routine choices between fixed alternatives in which case "his skill as a designer is not used and decays" (Rosenbrock 1977).

The same underlying systems design philosophy is now evident across most areas of intellectual activity. The outcome could be an abject dependence on systems and an inability to "think for ourselves." However, we still have a historical window which may well be closing but which might still allow for the design of systems in a symbiotic manner to make the best use of human attributes together with those of the system.

Half a century ago the Turing Test was devised to distinguish between human beings and machines. All around the world today we see examples of humans behaving more like machines and machines more like human beings. The development is in the form of a convergence whereas what is required is one based on symbiosis.

The Search for Alternatives

Parody becomes reality

In the BBC comedy series *Little Britain*, the character Carol is a bored and indifferent bank employee. When a customer asks for a £2,000 loan she types in a few figures and declares smugly: "Computer says no." Becoming increasingly anxious the customer makes a number of suggestions including a smaller loan and meeting the Manager. Getting the same response the customer makes a final attempt saying "Is there anything I can do?" Carol whispers to the computer and repeats "computer says no."

All of this is so resonates with the public's experience that there is a now brisk market for badges, fridge magnets, key-rings and cartoons bearing the slogan "computer says no." You can even get a ring tone for your mobile declaring it. In the parody Carol at least speaks to the customer but the reality can be much more alarming.

When a Rochdale resident had no response whatsoever to three urgent e-mail messages to the Council's Planning Department objecting to the erection of a structure, he eventually established that the messages had been screened out by Rochdale's anti-porn software due to his inclusion of the dreaded word "erection" (press report 2006). The computer had said "no" and the plans were passed before the protest could be considered as the system was devoid of the contextual understanding that a human being would have applied. Such experiences are now becoming common place even as IT equipment manufacturers proudly proclaim in adverts that their products help you "Take back control."

On your bike

The nature of technological change in its current form is that propositional knowledge becomes more significant than tacit knowledge. This results in "know that" being more important that "know how."

Tacit knowledge comes from "learning by doing" and results in the ability to judge situations based on experience. Propositional knowledge is based more on analysis and calculation. Within the

From judgement to calculation

human centred tradition, a symbiosis of the two and a creative interaction of them is essential. This is particularly true in the case of skilled activities.

The nature of tacit knowledge is that (to quote Polanyi): "There are things we know but cannot tell." In his seminal paper he continues:

> "I can say I know how to ride a bicycle or how to swim but it does not mean that I can tell how I managed to keep my balance on a bike or keep afloat when swimming. I may not have the slightest idea of how I do this or even an entirely wrong or grossly imperfect idea of it and yet can go on cycling and swimming merrily."

He points out that there are two kinds of knowledge which invariably enter jointly into any act of knowing a complex entity. There is firstly knowing a thing by attending to it. In that way we attend to the entity as a whole. And secondly there is knowing a thing by relying on our awareness of its purpose of attending to an entity to which it contributes. A detailed explanation of this is given by Polanyi himself (Polanyi 1962).

USE–ABUSE

One of the key strands of the debate about human centred systems in the UK arose not so much in academic circles as in the industrial context of Lucas Aerospace. The company employed some 18,000 skilled craftsmen, prototype fitters, engineers, metallurgists, control systems engineers, scientists and laboratory staff. In the early seventies the company was one of the world's largest manufacturers of aerospace actuators, generators, systems and auxiliary items.

It was clear that the company was embarking on a rationalisation strategy and it eventually emerged that some 4,000 of these world class technologists were facing unemployment. Several leading members of trade unions were engaged in debates in the continuing discussion from the sixties on the role S&T in society. These discussions went far beyond the use/abuse model and questioned the nature of S&T itself.

There was a vigorous discussion about the gap between the potential of (S&T) and its reality. Furthermore, there was a questioning of the assumption that science – in its own terms at least – had come to monopolise the notion of the rational and could therefore be counter-posed with irrationality and suspicion. Indeed it came to be seen as a means by which irrationality could be exercised. In discussions and exchanges of correspondence with organisations such as the British Society for Social Responsibility in Science through to academics in the USA, it gradually began to be realised that far from being neutral, S&T actually reflected sets of values causing us to speak in terms of the control of nature, the exploitation of natural resources and the manipulation of data.

ONE BEST WAY?

It was clear that within S&T there is the notion of the "one best way." However, viewing them as part of culture which produced different music, different literature and different artefacts, why should there not be differing forms of S&T? Furthermore there was an increasing realisation that S&T had embodied within it many of the assumptions of the society giving rise to it.

Space will not permit a detailed exploration of these extraordinary developments. Suffice it to say that the workforce produced a plan for what they called "Socially Useful, Environmentally Desirable Production." They produced and demonstrated a road/rail vehicle, prototypes of city cars and they designed and produced a range of medical products all as an alternative to structural unemployment. There were also a variety of products proposed for third world countries. In discussions dealing with how these products would be produced, it was suggested that producing these in the usual Tayloristic, alienating fashion would be unacceptable and so there arose in parallel a searching and probing discussion about the notion of human centred systems which would celebrate human skill and ingenuity

From judgement to calculation

rather than subordinate the human being to the machine or system.

In the discussions which led to the widely acclaimed Lucas Workers' Plan (Cooley 1991), there needed to be practical examples so that the polarised options of development could be recognised. That is, whether the process should be total automation and machine based systems or those which would build on human skill and ingenuity.

The EEC sponsored major research programme with research institutes and private companies in Denmark, Germany and the UK to produce a human centred system and the positive results of this are reported elsewhere (Cooley 1993).

Telechirics – high tech, high touch

Another practical example arose from the design need to produce a submersible vehicle capable of carrying out repairs in hazardous offshore environments. Initial considerations of a highly automated device indicated the huge computing and feedback capabilities necessary if humans were to be excluded from the process.

It was recognised that a telechiric device could work in a remote and hazardous environment but provide feedback – audio, tactile and visual – to skilled operators in a safe environment. Such devices were already in use in other hazardous environments such as nuclear power. Thus telechiric devices became one of the product proposals in the Lucas Plan and emphasis was laid upon the wider application of such environments, not least in the medical field. In all cases the systems were designed such as to celebrate and enhance the skill and ability to judge of the human beings involved.

Look and feel

In the case of surgery, some of the sensationalist press headlines refer to "robotic surgeons." In fact the reality is that some of these systems would enhance the skill of the surgeon rather than diminish it. An example is in the field of minimal invasive surgery. These systems provide enhanced dexterity precision and control which may be applied to many surgical procedures currently performed using standard laparoscopic techniques. In fact the systems now reported, succeed in providing the surgeons with all the clinical and technical capabilities of open surgery whilst enabling them to operate through tiny incisions.

As one of the companies producing such systems point out, it succeeds in maintaining the same "look and feel" as that of open surgery. The surgeon is provided with a "tool" to enhance and extend his or her skill whilst the patient may experience a whole range of improved outcomes, e.g. reduced trauma to the body, shorter hospital stay, less scarring and improved cosmesis. It is the judgment of the skilled surgeon that drives the system, not the technology.

CAVALIER DISREGARD

As *AI & Society* celebrates its 21st birthday, it is gratifying to see the emergence of some systems displaying many of the symbiotic attributes the journal has been espousing. Alas, the dominant tendency is still to confer life on systems whilst diminishing human involvement. Designers do so with cavalier disregard for potential human competence. Quantitative comparisons of human and systems capabilities are questionable and they do not compare like with like. However it is sobering on occasion to reflect upon the ball park comparison. Thus Cherniak (Cherniak 1988) suggests that the massive battle management software of the Strategic Defence Initiative is "at least a 100 times smaller than the estimated size of the mind's programme."

NETWORKS

Human and technology networks can encourage and stimulate people to be innovative and creative. To encourage people to think in these terms, we need a form of enterprise culture. However,

From judgement to calculation

universities and conventional secondary schools disregard such attributes because many are not predictable, repeatable or quantifiable. From a democratic standpoint, we need to redirect S&T because more and more of our citizens are opposed to its present form and to those who own and control it.

A recent survey of EU citizens shows that if you ask them whom they can believe when informed about issues such as bio-engineering and genetic modification, only about 21% believe that you can accept what the multinationals tell you which suggests to me that there are still a lot of trusting people out there. Then if you ask them what about universities, only 28% say that you can believe what the universities say because they are frequently apologists for the big companies.

However, if you ask them "Can you believe what Greenpeace tells you?" 54% will say "Yes."

Now this survey is a very important warning for us. If we have lost the trust of our citizens its no use pleading that they cannot or have not understood, for it is our fault for failing to communicate adequately. There are ways of communicating if we really want without making a virtue out of complexity.

KINDRED SPIRITS

Challenging the given orthodoxy is a precarious and lonely affair. It is therefore important to build up and participate in a supportive network of kindred spirits. This may take many forms, one example is the Institute Without Walls set up by *AI & Society* colleagues. The exchange of ideas and the development of collaborative projects are all important.

The support of funding bodies was likewise important with the Greater London Enterprise Board gaining EU research funding for Esprit 1217 – to design and build and demonstrate a human centred manufacturing system. Funding was also made available by the EU FAST project – to set up a team of experts from EU member states which would produce a report.

The Search for Alternatives

The ensuing report was entitled 'European Competitiveness in the twenty-first century: the integration of Work, Culture and Technology.' It was part of the FAST proposal for an R&D programme on 'Human Work in Advanced Technological Environments.' The report provided practical examples of human enhancing systems and called for an industrial and cultural renaissance. It advocated that new forms of education should facilitate the transmission of a culture valuing proactive, sensitive and creative human beings.

In 1990, the EU commissioned and published 26 reports in its Anthropocentric Production Systems (APS) research papers series. Several of these were based on an analysis of the potential of APS for individual member states.

CHERISH SKILL AND JUDGMENT

During the formulation of the original ideas, the International Metalworkers' Federation held a conference in 1984 and hosted a presentation by the author entitled 'Technology, Unions and Human Needs.' The presentation, subsequently published as a 58 page report in 11 languages including Finnish and Japanese, was circulated to the Federation's members worldwide.

Publicity for these ideas at the more popular level was also important as it is spurious to talk about a democratic society if the public can not influence the manner in which technology is developing. In this context, the one hour television programme in the Channel 4 (London) *Equinox* series which was presented by the author, caused considerable interest as did a number of interviews and articles in the more popular press. TV Choice London produced an educational video 'Factory of the Future' explaining the application of human centred systems which valorize human skill and judgment.

From judgement to calculation

THE WRIGGLING WORM

Education – like democracy – can only be partially given and for the remainder, it must be taken. Indeed, taking it is part of the process itself. Some of those designing IT systems for education behave as though a body of knowledge can be downloaded on to a human brain. It is true that some of these systems are impressive and used as a tool to aid human learning they are, and will continue to be, of great significance. The range of options, images and supporting films and graphic animation can indeed be overwhelming. However, it should be noted that in many cases they come between us and the real world. They provide us with forms of second and third order reality and information.

This may be explained by a simple example. Any child can get an impressive range of support from the internet and learning systems but this form of knowledge is very different from that acquired by one who goes into their local wood, lifts up a stone, picks up a worm and feels it wriggling in the palm of his hand. To this tactile input may then be added contextual information – summer or winter? Farms in the background? Was there the scent and feel of damp soil or decaying leaves?

So I suggest that in education in coming years, we are going to acquire learning in developing situations where there will be the form of explicit knowledge, you acquire in a university, but of equal importance will be the implicit knowledge and the informal situations that really advise our lives. It is essential to understand that if we just proceed on this mechanistic basis, the mistakes we make will be truly profound and creative opportunities will be missed.

NATURAL SCIENCE?

We are frequently told that the best way we can proceed is within a rule-based system. This is absolutely extraordinary! As any active trade unionist knows, the way to stop anything in its tracks is to work to rule. It is all the things that we do outside the rule-

The Search for Alternatives

based system that keeps everything going.

As matters now stand, the given scientific methodology can only accept that a procedure is scientific if it displays the three predominant characteristics of the natural sciences: predictability, repeatability and mathematical quantifiability. These by definition preclude intuition, subjective judgment, tacit knowledge, dreams, imagination, heuristics, motivation and so I could go on. So instead of calling these the natural sciences, perhaps they should be renamed the unnatural sciences. There are other ways of knowing the world than by the scientific methodology.

Furthermore, when we talk of informating people rather than automating them we need to be clear that we are talking about information and not data. Transforming data into information requires situational understanding which the human can bring to bear. This information can then be so applied as to become knowledge which in turn is absorbed into a culture and thereby becomes wisdom (Cooley 2002).

THE MISTRESS EXPERIENCE

Reductionists have much to answer for. They have intimidated those who proceed on the basis of tacit knowledge. Even the giants of our civilisation were derided by them.

Thus we have Leonardo's spirited riposte: "They say that not having learning, I will not speak properly of that which I wish to elucidate. But do they not know that my subjects are to be better illustrated from experience than by yet more words? Experience, which has been the mistress of all those who wrote well and thus, as mistress, I will cite her in all cases" (Cooley 1991). The academic reductionists had even enacted a law to prevent master builders calling themselves a "master" because it may have been confused with the academic title "magister."

From judgement to calculation

PERFECT FLOWER OF GOOD MANNERS

As early as the thirteenth century, Doctors of Law were moved to protest formally at these academic titles being used by practical people whose structures and designs demonstrated competence of the highest order. Thus the separation between intellectual and manual work; between theory and practice, was being further consolidated at that stage and the title Dr Lathomorum was gradually eliminated. The world was already beginning to change at the time when the following epitaph could be written for the architect who constructed the nave and transepts of Saint Denis: "Here lies Pierre de Montreuil, a perfect flower of good manners, in this life a Doctor of Stones."

Significantly, following this period and in most the European languages there emerged the word DESIGN or its equivalent, coincident with the need to describe the occupational activity of designing. This is not to suggest that designing was a new activity, rather it indicated that designing was to be separated from doing and tacit knowledge separated from propositional knowledge (Cooley 1991).

LIBERATING HUMAN IMAGINATION

Within the human centred tradition, liberating human imagination is pivotal. This is true in the hardest of the sciences as it is in music or literature. Einstein said on one occasion "imagination is more important than knowledge." Furthermore, when pressed to reveal how he arrived at the theory of relativity, he is said to have responded: "When I was a child of 14 I asked myself what it might be like to ride on a beam of light and look back at the world."

In a wider sense, we need to emphasise all the splendid things that humans can do.

This is in contrast to the defect model which emphasises what they cannot do. The destructiveness of viewing humans in this manner is dramatically highlighted in the extraordinary passage in James Joyce's *Finnegans Wake* when he describes the purveyors of this negative approach as:

> "Sniffer of carrion, premature gravedigger, seeker of the nest of evil in the bosom of a good word, you, who sleep at our vigil and fast for our feast, you with your dislocated reason..." (Cooley 2005).

Confucius

This article has been wide ranging and will have raised a number of controversial issues. The references provide a framework in which to explore the ideas further. Some parts of it deal with cutting edge new technologies, yet it is gratifying to think that we can revert to Confucius to encapsulate these ideas so succinctly.

"I hear and I forget. I see and I remember. I do and I understand"

Now on this 21st birthday the journal can be proud of the impressive body of work it has nurtured. This augers well for the future development of systems which will be more caring of humanity and our precious planet.

References

Cherniak C (1988) Undebuggability and cognitive science. Commun Assoc Comput Mach 31(4):402–412

Cooley M (1991) Architect or bee?: the human price of technology. Chatto & Windus/The Hogarth Press, London. 2nd Impression 1991

Cooley M (1993) Skill and competence for the 21st century. PROC: IITD conference, Galway, April 1993

Cooley M (2002) Stimulus points: making effective use of IT in health. Workshop. Post Grad

Department. Brighton & Sussex Medical School 14.10.2002

Cooley M (2005) Re-Joyceing engineers. Artif Intell Soc 19:196–198

Dreyfus HL, Dreyfus SE (1986) Mind over machine. The Free Press 1986

From judgement to calculation

Joseph C (1999) Article, The Times, London 20.04.1999

Mazlish B (1967) The fourth discontinuity. Technol Cult 8(1):3–4

Mumford L (1963) Technics and civilisation. Harcourt Brace Jovanovich, New York, pp 13–15

Pearson K (1976) Computer power and human reason (cited in Weizenbaum J). WH Freeman & Co, San Francisco, p 25

Polanyi M (1962) Tacit knowing: its bearing on some problems of philosophy. Rev Mod Phys 34(4):601–616

Press report (2006) Report in Daily Mail 31.05.2006

Rapoport A (1963) Technological models of the minds. In: Sayre KM, Crosson FJ (eds) The modelling of the mind: computers and intelligence. The University of Notre Dame Press, pp 25–28

Rogers L (1999) Article, Sunday Times 18.04.1999, p 7

Rosenbrock HH (1977) The future of control. Automatica 13:389–392

Rosenbrock HH (1988) Engineering as an art. Artif Inell Soc 2:315–320

Rosenbrock HH (1990) Machines with a purpose. Oxford University Press, Oxford, pp 122–124. (See also Book Review in AI & Society vol 5, no.1)

Rosenbrock HH (2002) USSHER cited in ''A Gallimaufry of Quaint Conceits''. Control Systems Centre, UMIST

Weizenbaum (1976) Computer power and human reason. WH Freeman & Co., San Francisco, p 25

First published:
AI & SOCIETY, 21:395-409 (2007)

Also available by Mike Cooley:

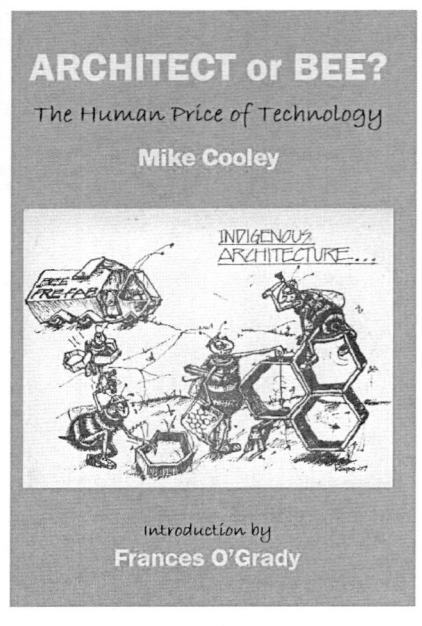

Architect or Bee?
The Human Price of Technology

Mike Cooley's pioneering introduction to socially useful production and human-centred manufacturing systems, with a new foreword by the General Secretary of the TUC. Years ahead of its time, Architect or Bee? is essential reading for all those seeking a democratic alternative to the politics of austerity.

Price: £10.99
194 pages | Paperback
ISBN: 978 0 85124 8493

Delinquent Genius
The Strange Affair of Man and His Technology

"Delinquent Genius is simply brimming with insights. It traces the sources of technology and its application. It is, above all else, a brilliant account of a dangerous hubris which can lead to that which is instrumental becoming a source, a dangerous source of domination, of passive rather than active existence..."

Price: £11.99
248 Pages | Paperback
ISBN 978 085124 8783

www.spokesmanbooks.com